GW00482778

RESTAURANT
AND
COMMERCIAL COOKING
FIRE INVESTIGATIONS

By
JAMES F. VALENTINE, JR.

Printed in the United States
ACTION GRAPHICS
424 Gibbsboro Road, Lindenwold, NJ 08021
Library of Congress Registration No: TX 7-366-086

Second Edition

CONTENTS

PREFACE

This book is written to assist an origin and cause investigator in the examination of a commercial cooking fire. A Certified Fire Investigator (CFI) or Certified Fire Explosion Investigator (CFEI) or other professional certified in the investigation process should only use this book as a guide.

The National Fire Protection Association, Standard 921, "*Guide for Fire Investigations*" and "*Kirk's Fire Investigation*" are to be utilized, along with other authoritative documents in reaching an area or point of origin of a commercial cooking fire and ensuing fire progression.

The Manufacturers Manual, UL Listing and the National Fire Protections Associations Standards 17, "*Dry Chemical Systems*", NFPA 17A, "*Wet Chemical Systems*", NFPA 96, "*Ventilation*"; and "*Fire Protection for Commercial Cooking Operations*" *should also be utilized during* the investigation process.

Note: This book is designed as a guide for professionals in the field. Nothing can replace fire experience and this book is not intended to replace the knowledge that comes with actual field-work. Use this book as a guide and always use an experienced investigator.

ACKNOWLEDGMENTS

I wish to thank Chief Joseph Rizzo of the Philadelphia Fire Department and David Demers, PE for their belief in my ability, as well as the late Warren Driscoll of Chemetron Fire Systems for my start in the suppression business.

To Assistant Chief Charles Wells and the officers and firefighters of the Lindenwold Fire Department, Camden County, New Jersey, who I am proud to call my brothers.

To my wife Dianne, who has always been by my side with her support, and my son Sean and daughter Carly, who have given me so much happiness and pride. Finally, to my parents, Jim and Marge.

DEDICATION

I wish to dedicate this book to the memory of Camden County Fire Marshall and Mt. Ephraim Fire Department Deputy Chief John West, an excellent fire investigator and fire fighter, who made the ultimate sacrifice while trying to rescue a child on July 4, 2002, Gloucester City, New Jersey.

CHAPTER 1

Introduction

There are numerous causes for fires during commercial cooking operations, such as human error, mechanical failures or improper maintenance of the ventilation and fire protection systems.

Commercial cooking flame flare-ups are a common occurrence that by restaurant standards are simply extinguished with a wet towel, milk, flour or cornstarch. Flare-ups can occur during the simple flipping of a burger or the more difficult task of flam baying of a dish.

Failure of a cooking device such as a fryer, can cause a runaway condition of the unit to occur. With the increase of cooking temperature being created auto-ignition can occur. By definition "auto ignition" is the ignition of oil without a spark or open flame being present. Ignition of the grease by an open flame can also occur during the malfunction of a cooking appliance.

Due to the amount of meat product being cooked or the amount of frying being done, grease-laden vapors begin to collect throughout the cooking appliance line and ventilation system. During cooking operations, the warm air venting throughout the ventilation system allows the accumulated grease to liquefy, which in turn allows vapors to be given off. Grease itself is a combustible material with a flash point of over 100 degrees Fahrenheit. When grease vapors are within their lower and upper combustible limits ignition will occur and the ensuing fire will progress up and into the ventilation system and drawn out the exhaust fan.

The intent of the Standard and Codes is to encapsulate the fire until the suppression system activates and extinguishes the fire. However, when the standards are violated or improper cleaning of the ventilation system or the inspection of the suppression system is incorrectly done, the encapsulated fire will extend from within the ventilation system and ignite the combustible material of the building construction causing a structural fire.

The following chapters will discuss the proper installation, service and inspection of the ventilation and suppression systems.

CHAPTER 2

Ventilation of Commercial Cooking Operations

The main purpose of the cooking ventilation system is to encapsulate the cooking vapors that are being given off from the cooking appliances. In the case of a fire occurrence, the same ventilation system must be able to withstand the temperatures being given off from the fire spreading within the ventilation system and to encapsulate that fire without ignition of any structure members surrounding the ventilation system. The fire must stay within the ventilation hood and ducts until the suppression system can activate, either by automatic or manual means to extinguish the fire.

The NFPA 96, "*Standard for Ventilation Control and Fire Protection of Commercial Cooking Operations*", is the standard that "*shall provide the minimum fire safety requirements (preventative and operative) related to the design, installation, operation, inspection and maintenance of all public and private cooking operations.*"

As stated under A.1.1.1.,"*These requirements include, but are not limited to, all manner of cooking equipment, exhaust hoods, grease removal devices, exhaust ductwork, exhaust fans, dampers, fire extinguishing equipment, and all other auxiliary or ancillary components or systems that are involved in the capture, containment, and control of grease-laden cooking effluent.*"

The most important statement in NFPA 96 is: "*THIS STANDARD CANNOT PROVIDE SAFE DESIGN AND OPERATION IF PARTS OF IT ARE NOT ENFORCED OR ARE ARBITRARILY DELETED IN ANY APPLICATION.*"

In other words, the standard must be adhered to in its entirety and cannot be deviated from unless the components being utilized have their own Listing of Installation. The standard will only be as strong as its weakest link.

CHAPTER 3

NFPA 96 Definitions

The following two chapters contain a copy of the definitions and requirements of the NFPA 96. Comments are made by the author and intended for purposes of interpretation and explanation, only.

3.1 General
The definitions contained in this chapter shall apply to the terms used in this standard. Where terms are not included, common usage of the terms shall apply.

3.2.1* Approved. Acceptable to the authority having jurisdiction.

3.2.2* Authority Having Jurisdiction (AHJ). An organization, office or individual responsible for enforcing the requirements of a code or standard, or for approving equipment, materials, an installation or a procedure.

3.2.3 Labeled. Equipment or materials to which has been attached a label, symbol, or other identifying mark of an organization that is acceptable to the authority having jurisdiction and concerned with product evaluation, that maintains periodic inspection of production of labeled equipment or materials, and by whose labeling the manufacturer indicates compliance with appropriate standards or performance in a specified manner.

3.2.4* Listed. Equipment, materials or services included in a list published by an organization that is acceptable to the authority having jurisdiction and concerned with evaluation of products or services, that maintains periodic inspection of production of listed equipment or materials or periodic evaluation of services, and whose listing states that either the equipment, material or service meets appropriate designated standards or has been tested and found suitable for a purpose.

3.2.5 Shall. Indicates a mandatory requirement.

3.2.6 Should. Indicates a recommendation or that which is advised but not required.

3.2.7 Standard. A document, the main text of which contains only mandatory provisions using the word "shall" to indicate requirements and which is in a form generally suitable for mandatory reference by another standard or code or for adoption into law. Non-mandatory provisions shall be located in an appendix or annex, footnote or fine-print note and are not to be considered a part of the requirements of a standard.

3.3.General Definitions.

3.3.1 Access Panel. A closure device used to cover an opening into a duct, an enclosure, equipment or an appurtenance.

3.3.2 Air Intake. An opening in a building's envelope whose purpose is to allow outside air to be drawn into the structure to replace inside air that is removed by exhaust systems or to improve the quality of the inside air by providing a source of air having a lower concentration of odors, suspended particles, or heating content.

3.3.3 Air Pollution Control Devices. Equipment and devices used for the purpose of cleaning air passing through them or by them in such a manner as to reduce or remove the impurities contained therein.

3.3.4* Appliance Flue Outlet. The opening or openings in a cooking device where vapors, combustion gases, or both leave the cooking device.

3.3.5 Appurtenance. An accessory or a subordinate part that enables the primary device to perform or improve its intended function.

3.3.6 Automatic. That which provides a function without the necessity of human intervention. [101:3.3].

3.3.7 Baffle Plate. An object placed in or near an appliance to change the direction or to retard the flow of air, air fuel mixtures, or flue gases.

3.3.8 Broiler.

3.3.8.1 High Broiler. See Upright Broiler.

3.3.8.2 Salamander Broiler. See Upright Broiler.

3.3.8.3 Upright Broiler. An appliance used in the preparation of food whereby foods are exposed to intense radiant heat, and perhaps to convective heat, with the food or the food and the radiant source not limited to a horizontal mode.

3.3.9*Certified. A formally stated recognition and approval of an acceptable level of competency, acceptable to the AHJ.

3.3.10 Classified. Products or materials of a specific group category that are constructed, inspected, tested and subsequently reinspected in accordance with an established set of requirements. The classification process is performed by an organization acceptable to the authority having jurisdiction. [80:1.4].

3.3.11 Clearly Identified. Capable of being recognized by a person of normal vision without causing uncertainty and indecisiveness about the location or operating process of identified item.

3.3.12* Construction.

3.3.12.1 Closed Combustible Construction. Combustible building construction including walls, structural framing, roofs, roof ceilings, floors and floor-ceiling assemblies continuously enclosing a grease duct on four sides where one or more sides are protected.

3.3.12.2 Open Combustible Construction. Combustible building construction including wall, structural framing, roof, roof ceiling, floor and floor-ceiling assemblies adjacent to a grease duct on three or fewer sides where one or more sides are protected.

3.3.13* Continuous Weld. A metal-joining method that produces a product without visible interruption or variation in quality.

3.3.14 Damper. A valve or plate for controlling draft or flow of gases including air.

3.3.15 Detection Devices. Electrical, pneumatic, thermal, mechanical or optical sensing instruments, or sub components of such instruments, whose purpose is to cause an automatic action upon the occurrence of some preselected event.

3.3.16 Dips. Depression or cuplike places in horizontal duct runs in which liquids could accumulate.

3.3.17 Discharge. The final portion of a duct or pipe where the product being conveyed is emptied or released from confinement; the termination point of the pipe or duct.

3.3.18 Duct Termination. The final or intended end-portion of a duct system that is designed and functions to fulfill the obligations of the system in a satisfactory manner.

3.3.19 Ducts (or Duct System). A continuous passageway for the transmission of air and vapors that, in addition to the containment components themselves, might include duct fittings, dampers, plenums, and/or other items or air-handling equipment.

3.3.19.1 Bleed Air Duct. An intake duct system, designed to input air to maintain system balance.

3.3.19.2 Grease Ducts. A containment system for the transportation of air and grease vapors that is designed and installed to reduce the possibility of the accumulation of combustible condensation and the occurrence of damage if a fire occurs within the system.

3.3.20 Easily Accessible. Within comfortable reach, with limited dependence on mechanical devices, extensions, or assistance.

3.3.21 Enclosure.

3.3.21.1 Continuous Enclosure. A recognized architectural or mechanical component of s building having a fire resistance rating as required for the structure and whose purpose is to enclose the vapor removal duct for its full length to its termination point outside the structure without any portion of the enclosure having a fire resistance rating less than the required value.

3.3.21.2 Grease Duct Enclosure.

3.3.21.2.1 Factory-Built Grease Duct Enclosures. A listed factory-built grease duct system for reduced clearances to combustibles and as an alternative to a duct with its fire-rated enclosure.

3.3.21.2.2 Field-Applied Grease Duct Enclosure. A listed system evaluated for reduced clearances to combustibles and as an alternative to a duct with its fire-rated enclosure.

3.3.22 Equipment.

3.3.22.1 Fire-Extinguishing Equipment. Automatic fire-extinguishing systems and portable fire extinguishers provided for the protection of grease removal devices, hoods, duct systems, and cooking equipment and listed for such use.

3.3.22.2* Solid Fuel Cooking Equipment. Cooking equipment that utilizes solid fuel.

3.3.23 Filter.

3.3.23.1* Grease Filter. A removable component of the grease removal system designed to capture grease and direct it to a safe collection point.

3.3.23.2* Mesh-Type Filer. A general purpose air filter not listed for or intended for grease applications.

3.3.24 Fire Resistance Rating. The time, in minutes or hours, that materials or assemblies have withstood a fire exposure as established in accordance with the test procedures of NFPA 251, Standard Methods of Tests of Fire Endurance of Building Construction and Materials. [150:1.4]

3.3.25 Fire Walls. A wall separating building or subdividing a building to prevent the spread of the fire and having a fire resistance rating and structural ability.

3.3.26 Fume Incinerators. Devices utilizing intense heat or fire to break down and/or oxidize vapors and odors contained in gases or air being exhausted into the atmosphere.

3.3.27 Fusible Link. A form of fixed temperature heat detecting device sometimes employed to restrain the operation of an electrical or mechanical control until its designed temperature is reached.

3.3.28* Grease. Rendered animal fat, vegetable shortening, and other such oily matter used for the purposes of and resulting from cooking and/or preparing foods.

3.3.29 Grease Removal Devices. A system of components designed for and intended to process vapors, gases and/or air as it is drawn through such devices by collecting the airborne grease particles and concentrating them for further action at some future time, leaving the existing air with a lower amount of combustible matter.

3.3.30 Greasetight. Constructed and performing in such a manner as not to permit the passage of any grease under normal cooking conditions.

3.3.31 High Limit Control Device. An operating device installed and serving as an integral component of a deep fat fryer that provides secondary limitation to the grease temperature by disconnecting the thermal energy input when the temperature limit is exceeded.

3.3.32* Hood. A device provided for a cooking appliance(s) to direct and capture grease-laden vapors and exhaust gases.

3.3.32.1. Fixed Baffle Hood. A Listed unitary exhaust hood design where the grease removal device is a non-removable assembly that contains an integral fire-activated water-wash fire-extinguishing system listed for this purpose.

3.3.33 Interconnected. Mutually assembled to another component in such a manner that the operation of one directly affects the other or that the contents of one specific duct system are allowed to encounter or contact the products being moved by another duct system.

3.3.34 Liquid tight. Constructed and performing in such a manner as not to permit the passage of any liquid at any temperature.

3.3.35* Material.

3.3.35.1 Combustible Material. A material capable of undergoing combustion.

3.3.35.2 Limited-Combustible Material. Refers to a building construction material not complying with the definition of non-combustible material that, in the form in which it is used, has a potential heat value not exceeding 3500 Btu/lb (8141 kJ/kg), where tested in accordance with NFPA 259 and includes (1) materials having a structural base of non-combustible material, with a surfacing not exceeding a thickness of in. (3.2 mm)that has a flame spread index not greater than 50; and (2) materials, in the form and thickness used, other than as described in (1), having neither a flame spread index greater than 25 nor evidence of continued progressive combustion, and of such composition that surfaces that would be exposed by cutting through the material on any plane would have neither a flame spread index greater than 25 nor evidence of continued progressive combustion, and of such composition that surfaces that would be exposed by cutting through the material on any plane would have neither a flame spread index greater than 25 nor evidence of continued progressive combustion. [50000:3.3]

3.3.35.3* non-combustible Material. A material not capable of supporting combustion.

3.3.36 Pitched. To be fixed or set at a desired angle or inclination.

3.3.37 Qualified. A competent and capable person or company that has met the requirements and training for a given field acceptable to the AHJ.

3.3.38 Recirculating Systems. Systems for control of smoke or grease-laden vapors from commercial cooking equipment that do not exhaust to the outside.

3.3.39 Removable. Capable of being transferred to another location with a limited application of effort and tools.

3.3.40 Replacement Air. Air deliberately brought into the structure, then specifically to the vicinity of either a combustion process or a mechanically or thermally forced exhausting device, to compensate for the vapor and/or gases being consumed or expelled.

3.3.41 Single Hazard Area. Where two or more hazards can be simultaneously involved in fire by reason of their proximity, as determined by the authority having jurisdiction.

3.3.42 Solid Cooking Fuel. Any solid, organic, consumable fuel such as briquettes, mesquite, hardwood or charcoal.

3.3.43 Solvent. A substance (usually liquid) capable of dissolving or dispersing another substance; a chemical component designed and used to convert solidified grease into a liquid or semi liquid state in order to facilitate a clearing operation.

3.3.44 Space.

3.3.44.1 Concealed Spaces. That portion(s) of a building behind walls, over suspended ceilings, in pipe chases, attics and in whose size might normally range from 44.45 mm (1 in.) stud spaces to 2.44 m (8 ft) interstitial truss spaces and that might contain combustible materials such as building structural members, thermal and/or electrical insulation, and ducting.

3.3.44.2 Confined Spaces. A space whose volume is less than 1.42m'/293 W (50 ft'/1000 Btu/hr) of the aggregate input rating of all appliances installed in that space. [211:3.3]

3.3.45 Spark Arrester. A device or method that minimizes the passage of airborne sparks and embers into a plenum, duct and flue.

3.3.46 Thermal Recovery Unit. A device or series of devices whose purpose is to reclaim only the heat content of air, vapors, gases or fluids that are being expelled through the exhaust system and to transfer the thermal energy so reclaimed to a location whereby a useful purpose can be served.

3.3.47* Trained. A person who has become proficient in performing a skill reliably and safely through instruction and practice/field experience acceptable to the AHJ.

3.3.48 Trap. A cuplike or U-shaped configuration located on the inside of a duct system component where liquids can accumulate.

CHAPTER 4

NFPA General Requirements

4.1 General

4.1.1 Cooking equipment used in processes producing smoke or grease-laden vapors shall be equipped with an exhaust system that complies with all the equipment and performance requirements of this standard.

4.1.1.1* Cooking equipment that has been listed in accordance with UL 197 or an equivalent standard for reduced emissions shall not be required to be provided with an exhaust system.

4.1.1.2 The listing evaluation of cooking equipment covered by 4.1.1.1 shall demonstrate that the grease discharge at the exhaust duct of a test hood placed over the appliance shall not exceed 5 mg/m^3 when operated with a total airflow of 0.236 cubic meters per second (500 cfm).

The text of 4.1.1 has been revised by a tentative interim amendment (TIA). See page 1.

4.1.2 All such equipment and its performance shall be maintained in accordance with the requirements of this standard during all periods of operation of the cooking equipment.

4.1.3 The following equipment shall be kept in good working condition:

(1) Cooking equipment
(2) Hoods
(3) Ducts (if applicable)
(4) Fans
(5) Fire extinguishing systems
(6) Special effluent or energy control equipment

4.1.3.1 Maintenance and repairs shall be performed on all components at intervals necessary to maintain these conditions.

4.1.4 All airflows shall be maintained.

4.1.5 The responsibility for inspection, maintenance, and cleanliness of the ventilation control and fire protection of the commercial cooking operations shall be the ultimate responsibility of the owner of the system provided that this responsibility has not been transferred in written form to a management company or other party.

COMMENT: What is meant by responsibility? It means that the owner or his representative must make sure that the ventilation system is cleaned and the suppression system is inspected in accordance with the manufacturer's listing. How does one meet this responsibility? By having a professional, qualified individual or company come into the establishment and perform the required service and/or inspection. It does not mean that the owner must be qualified himself. The owner as does the AHJ has to rely on the qualified company or individual who is performing the required service and/or inspection to properly perform their duties in accordance with code, standard and the listings of the manufacturer.

4.1.6* All solid fuel cooking equipment shall comply with the requirements of Chapter 14. (See comments under Chapter 14)

4.1.7 Multi tenant applications shall require the concerted cooperation of design, installation, operation and maintenance responsibilities by tenants and by the building owner.

4.1.8 All interior surfaces of the exhaust system shall be accessible for cleaning and inspection purposes.

4.1.9* Cooking equipment used in fixed, mobile or temporary concessions, such as trucks, buses, trailers, pavilions, tents, or any form of roofed enclosure, shall comply with this standard.

4.1* Clearance.

4.2.1 Where enclosures are not required, hoods, grease removal devices, exhaust fans and ducts shall have a clearance of at least 457 mm(18 in.)to combustible material, 76 mm (3 in.) to limited-combustible material, and 0 mm (0 in.) to non-combustible material.

4.2.2 Where a hood, duct, or grease removal device is listed for clearance less than those required in 4.2.1, the listing requirements shall be permitted.

4.2.3 Clearance Reduction.

4.2.3.1 Where a clearance reduction system consisting of 0.33 mm (0.013 in.) (28 gauge) sheet metal spaced out 25 mm (1 in.) on non-combustible spacers is provided, there shall be a minimum of 229 mm (9 in.) clearance to combustible material.

4.2.3.2 Where a clearance reduction system consisting of 0.69 mm (0.027 in.) (22 gauge) sheet metal on 25 mm (1 in.) mineral wool batts or ceramic fiber blanket reinforced with mesh or equivalent spaced out 25 mm (1 in.) on non-combustible spacers is provided, there shall be a minimum of 76 mm (3in.) clearance to combustible material.

4.2.3.3 Zero clearance to limited-combustible materials shall be permitted where protected by metal lath and plaster, ceramic tile, quarry tile, other non-combustible materials or assembly of non-combustible materials, or materials and products that are listed for the purpose of reducing clearance.

4.2.4 Clearance Integrity.

4.2.4.1 In the event of damage, the material or product shall be repaired and restored to meet its intended listing or clearance requirements and shall be acceptable to the authority having jurisdiction.

4.2.4.2* In the event of a fire within a kitchen exhaust system, the duct and its enclosure (rated shaft, factory-built grease duct enclosure, or field-applied grease duct enclosure) shall be inspected by qualified personnel to determine whether the duct and protection method are structurally sound, capable of maintaining their fire protection function, and in compliance with this standard for continued operation.

4.2.4.3 Protection shall be provided on the wall from the bottom of the hood to the floor, or to the top of the non-combustible material extending to the floor, to the same level as required in 4.2.1.

4.4.4 The protection methods for ducts to reduce clearance shall be applied to the combustible or limited-combustible construction, not to the duct itself.

4.3 Field-Applied and Factory-Built Grease Duct Enclosures.

4.3.1 Field-applied grease duct enclosures and factory-built enclosures and factory-built grease duct enclosures shall be listed in accordance with UL 2221, *Standard for Tests of Fire Resistive Grease Duct Enclosure Assemblies*, or equivalent standard and installed in accordance with the manufacturer's instructions and the listing requirements.

4.3.2 Field-applied grease duct enclosures and factory-built grease duct enclosures shall demonstrate that they provide mechanical and structural integrity, resiliency, and stability when subjected to expected building environmental conditions, duct movements under general operating conditions and duct movement due to fire conditions.

4.3.3 The specifications of material, gauge, and construction of the duct used in the testing and listing of field-applied grease duct enclosures and factory-built grease duct enclosures shall be included as minimum requirements in their listing and installation documentation.

4.3.4 Clearance Options for Field-Applied and Factory-Built Grease Duct Enclosures. The following clearance options for which field-applied grease duct enclosures and factory-built grease duct enclosures have been successfully evaluated shall be clearly identified in their listing and installation documentation and on their label:

(1) Open combustible clearance at manufacturer's requested dimensions
(2) Closed combustible clearance at manufacturer's requested dimensions, with or without specified ventilation
(3) Rated shaft clearance at manufacturer's requested dimensions, with or without specified ventilation.

4.4 Building and Structural Duct Contact

4.4.1 A duct shall be permitted to contact non-combustible floors, interior walls and other non-combustible structures or supports, but it shall not be in contact for more than 50 percent of its surface area per each lineal foot of contact length.

4.4.2 Where duct contact must exceed the requirements of 4.4.1, the duct shall be protected from corrosion.

4.4.3 Where the duct is protected with a material or product listed for the purpose of reducing clearance to zero, the duct shall be permitted to exceed the contact limits of 4.4.1 without additional corrosion protection.

4.5 Duct Clearances to Enclosures. Clearances between the duct and interior surfaces of enclosures shall meet the requirements of Section 4.2.

4.6 Drawings. A drawing(s) of the exhaust system installation along with a copy of operating instructions for sub-assemblies and components used in the exhaust system, including electrical schematics, shall be kept on the premises.

4.7 Authority Having Jurisdiction Notification. If required by the authority having jurisdiction, notification in writing shall be given of any alteration, replacement, or relocation of any exhaust or extinguishing system or part thereof or cooking requirement.

CHAPTER 5

Hoods

5.1 Construction

5.1.1 The hood or that portion of a primary collection means designed for collecting cooking vapors and residues shall be constructed of and be supported by steel not less than 1.09 mm (0.043 in.) (No. 18 MSG) in thickness, stainless steel not less than 0.94 mm (0.037 in.) (No. 20 MSG) in thickness, or other approved material of equivalent strength and fire and corrosion resistance.

5.1.2 All seams, joints and penetrations of the hood enclosure that direct and capture grease-laden vapors and exhaust gases shall have a liquid tight continuous external weld to the hood's lower outermost perimeter.

COMMENT: One of the problems with hoods during a fire is the lack of liquid tight welds. Without containment the fire will escape.

5.1.3 Seams, joints, and penetrations of the hood shall be permitted to be internally welded, provided that the weld is formed smooth or ground smooth, so as to not trap grease and is clearable.

5.1.4* Internal hood joints, seams, filter support frames, and appurtenances attached inside the hood shall be sealed or otherwise made grease tight.

5.1.5 Penetrations shall be permitted to be sealed by devices that are listed for such use and whose presence does not detract from the hood's or duct's structural integrity.

5.1.6 Listed exhaust hoods with or without exhaust dampers shall be permitted to be constructed of materials required by the listing.

5.1.7 Listed exhaust hoods with or without exhaust dampers shall be permitted to be assembled in accordance with the listing requirements.

5.1.8 Eyebrow-Type Hoods.

5.1.8.1 Eyebrow-type hoods over gas or electric ovens shall be permitted to have a duct constructed as required in Chapter 7 from the oven flue(s) connected to the hood canopy upstream of the exhaust plenum as shown in Figure 5.1.8.1.

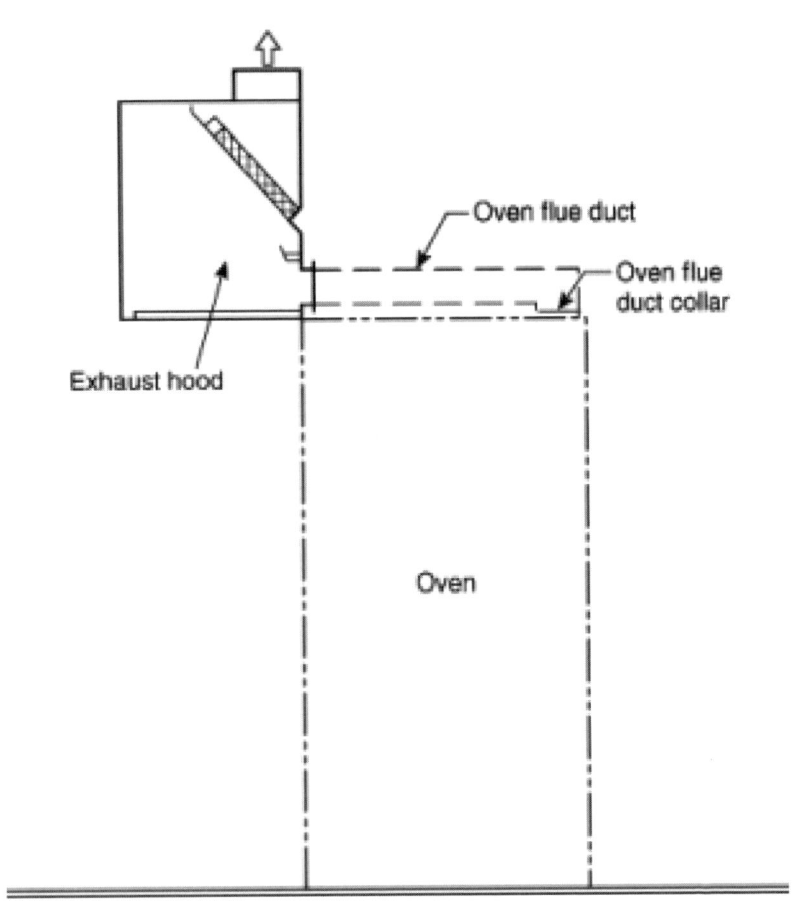

FIGURE 5.1.8.1
Typical Section of Eyebrow-Type Hood.

5.1.8.2 The duct connecting the oven flue(s) to the hood canopy shall be connected with a continuous weld or have a duct-to-duct connection.

5.1.9 Insulation materials other than electrical insulation shall have a flame spread rating of 25 or less when tested in accordance with UL 723.

5.1.10 Adhesives or cements used in the installation of insulating materials shall comply with the requirements of 5.1.9, when tested with the specific insulating material.

5.1.11 Penetration shall be sealed with listed devices in accordance with the requirements of 5.1.12.

5.1.12 Devices that require penetration of the hood, such as pipe and conduit penetration fittings and fasteners, shall be listed in accordance with UL 1978.

COMMENT: The breach into the hood by the suppression system without resealing the hole drilled for the pipe to pass through is something that is seen time and time again, mostly in dry chemical installations.

5.2 Hood Size. Hoods shall be sized and configured to provide for the capture and removal of grease-laden vapors. (See8.2.2)

5.3 Exhaust Hood Assemblies with Integrated Supply Air Plenums

5.3.1 The construction and size of exhaust hood assemblies with integrated supply air plenums shall comply with the requirements of Sections 5.1 and 5.2.

5.3.2 The construction of the outer shell or the inner exhaust shell shall comply with Section 5.1.

5.3.3 Where the outer shell is welded, the inner shell shall be of grease tight construction.

5.3.4* Fire Dampers.

5.3.4.1 A fire-actuated damper shall be installed in the supply air plenum at each point where a supply air duct inlet or a supply air outlet penetrates the continuously welded shell of the assembly.

5.3.4.2 The fire damper shall be listed for such use or be part of a listed exhaust hood with or without exhaust damper.

5.3.4.3 The actuation device shall have a maximum temperature rating of 141EC (286EF).

CHAPTER 6

Grease Removal Devices in Hoods

6.1 Grease Removal Devices.

6.1.1 Listed grease filters, listed baffles, or other listed grease removal devices for use with commercial cooking equipment shall be provided.

6.1.2 Listed grease filters shall be tested in accordance with UL 1046.

6.1.3 Mesh filters shall not be used.

COMMENT: Mesh filters will collect grease

whereas baffle filters cause grease molecules to compress together, becoming heavier than air and thereby dropping out of the air as the air passes through the baffle filter. Mesh filters can contribute to the failure of a suppression system to extinguish a fire. The suppression system is designed to extinguish surface fires therefore the collection of grease within a mesh filter causes the fire to become a deep-seated fire.

6.2 Installation.

6.2.1 Separation Distance

6.2.1.1 The distance between the grease removal device and the cooking surface shall be as great as possible but not less than 457.2 mm (16 in.)

6.2.1.2 Where grease removal devices are used in conjunction with charcoal or charcoal-type broilers, including gas or electrically heated charbroilers, a minimum vertical distance of 1.22 m (4 ft) shall be maintained between the lower edge of the grease removal device and the cooking surface.

6.2.1.3 For cooking equipment without exposed flame and where flue gases bypass grease removal devices, the minimum vertical distance shall be permitted to be reduced to not less than 152.4 mm (6 in.)

6.2.1.4 Where a grease removal device is listed for separation distances less than those required in 6.2.1.1 and 6.2.1.2, the listing requirements shall be permitted.

6.2.1.5 Grease removal devices supplied as part of listed hood assemblies shall be installed in accordance with the terms of the listing and the manufacturer's instructions.

6.2.2 Grease Removal Device Protection

6.2.2.1* Grease removal devices shall be protected from combustion gas outlets and from direct flame impingement occurring during normal operation of cooking appliances producing high flue gas temperatures, where the distance between the grease removal device and the appliance flue outlet (heat source) is less than 457.2 mm (18 in.).

6.2.2.2 This protection shall be permitted to be accomplished by the installation of a steel or stainless steel baffle plate between the heat source and the grease removal device.

6.2.2.3 The baffle plate shall be sized and located so that flames or combustion gases travel a distance not less than 457.2 mm (18 in.) from the heat source to the grease removal device.

6.2.2.4 The baffle shall be located not less than 152.4 mm (6 in.) from the grease removal devices.

COMMENT: This type of situation is usually associated with upright char broilers.

6.2.3 Grease Filters

6.2.3.1 Grease filters shall be listed and constructed of steel or listed equivalent material.

6.2.3.2 Grease filters shall be of rigid construction that will not distort or crush under normal operation, handling and cleaning conditions.

6.2.3.3 Grease filters shall be arranged so that all exhaust air passes through the grease filters.

6.2.3.4 Grease filters shall be easily accessible and removable for cleaning.

6.2.3.5 Grease filters shall be installed at an angle not less than 45 degrees from the horizontal.

6.2.4 Grease Drip Trays.

6.2.4.1 Grease filters shall be equipped with a grease drip tray beneath their lower edges.

6.2.4.2 Grease drip trays shall be kept to the minimum size needed to collect gases.

6.2.4.3 Grease drip trays shall be pitched to drain into an enclosed metal container having a capacity not exceeding 3.785 L (1 gal.).

6.2.5 Grease Filter Orientation. Grease filters that require a specific orientation to drain grease shall be clearly so designated, or the hood shall be constructed so that filters cannot be installed in the wrong orientation.

CHAPTER 7

Exhaust Duct System

COMMENT: In the majority of the restaurant cooking fire investigations that I have performed the most destructive damage is due to the improper duct installation allowing for the encapsulated fire within the ductwork to escape and ignite the roof/attic area of the structure.

7.1.1 Ducts shall not pass through fire walls.

7.1.2* All ducts shall lead directly to the exterior of the building, so as not to unduly increase any fire hazard.

7.1.3 Duct systems shall not be interconnected with any other building ventilation or exhaust system.

7.1.4 All ducts shall be installed without forming dips or traps that might collect residues. In manifold (common duct) systems, the lowest end of the main duct shall be connected flush with the bottom of the branch duct.

7.1.5 Openings required for accessibility shall comply with Section 7.3.

7.1.6 A sign shall be placed on all access panels stating the following:

ACCESS PANEL - DO NOT OBSTRUCT

7.1.7 Listed grease ducts shall be installed in accordance with the terms of the listing and the manufacturer's instructions.

7.2 Clearance. Clearance between ducts and combustible materials shall be provided in accordance with the requirements of Section 4.2.

COMMENT: The improper clearance of combustible material to the ductwork is a major problem. I am not sure that this improper clearance is due to lack of knowledge or misinterpretation of this standard.

7.3 Openings

COMMENT: Access panels are crucial for the proper and complete cleaning of the duct system. Without access panels inaccessible areas of the ductwork will exist allowing for the accumulation of grease.

7.4.1.2 Where an opening of this size is not possible, openings large enough to permit thorough cleaning shall be provided at 3.7 m (12 ft) intervals.

7.4.1.3 Support systems for horizontal grease duct systems 609 mm (24 in.) and larger in any cross-sectional dimension shall be designed for the weight of the ductwork plus 363 kg (800 lb) at any point in the duct systems.

7.4.1.4 On non-listed ductwork, the edge of the opening shall be not less than 38.1 mm (1 in.) from all outside edges of the duct or welded seams.

7.4.2 Vertical Ducts.

7.4.2.1 On vertical ductwork where personnel entry is possible, access shall be provided at the top of the vertical riser to accommodate descent.

7.4.2.2 Where personnel entry is not possible, adequate access for cleaning shall be provided on each floor.

7.4.2.3 On non-listed ductwork, the edge of the opening shall be not less than 38.1 mm (1 in.) from all outside edges of the duct or welded seams.

7.4.3 Access Panels.

7.4.3.1 Access panels shall be of the same material and thickness as the duct.

7.4.3.2 Access panels shall have a gasket or sealant that is rated for 815.6EC (1500EF) and shall be greasetight.

7.4.3.3 Fasteners, such as bolts, weld studs, latches, or wing nuts, used to secure the access panels shall be carbon steel or stainless steel and shall not penetrate duct walls.

7.4.3.4 Listed grease duct access door assemblies (access panels) shall be installed in accordance with the terms of the listing and the manufacturer's instructions.

7.4.4 Protection of Openings.

7.4.4.1 Openings for installation, servicing, and inspection of listed fire protection system devices and for duct cleaning shall be provided in ducts and enclosures and shall conform to the requirements of Section 7.3 and 7.7.4.

7.4.4.2 Enclosure openings required to reach access panels in the ductwork shall be large enough for the removal of the access panel.

7.5 Other grease Ducts. Other grease ducts shall comply with the requirements of this section.

7.5.1* Materials. Ducts shall be constructed of and supported by carbon steel not less than 1.37 mm (0.054 in.) (No. 15 MSG) in thickness or stainless steel not less than 1.09 mm (0.043.in.) (No. 18 MSG) in thickness.

7.5.2 Installation.

7.5.2.1 All seams, joints, penetrations and duct-to-hood collar connections shall have a liquid tight continuous external weld.

7.5.2.2 Duct-to-hood connections as shown in Figure 7.5.2.2 shall not require a liquid tight continuous external weld.

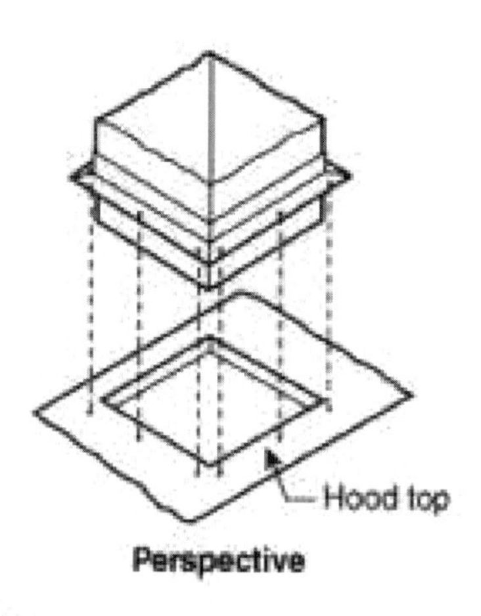

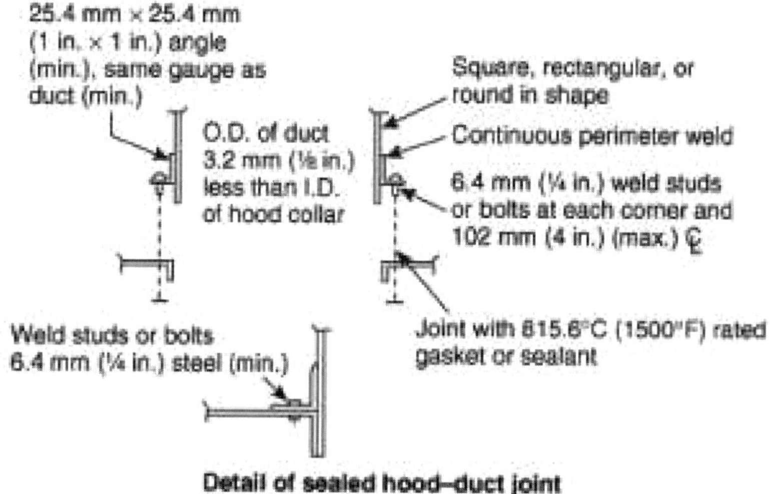

FIGURE 7.5.2.2
Permitted Duct-to-Hood Collar Connection.

7.5.2.3 Penetrations shall be permitted to be sealed by other listed devices that are tested to be grease tight and are evaluated under the same conditions of fire severity as the hood or enclosure of listed grease extractors and whose presence does not detract from the hood's or duct's structural integrity.

7.5.2.4 Internal welding shall be permitted, provided the joint is formed or ground smooth and is readily accessible for inspection.

7.5.2 Penetrations shall be sealed with listed devices in accordance with the requirements of 7.5.4.

7.5.4 Devices that require penetration of the ductwork, such as pipe and conduit penetration fittings and fasteners, shall be listed in accordance with UL 1978.

7.5.5 Duct Connections

7.5.5.1 Acceptable duct-to-duct connections shall be as follows

(1) Telescoping joint as shown in Figure 7.5.5.1(a)

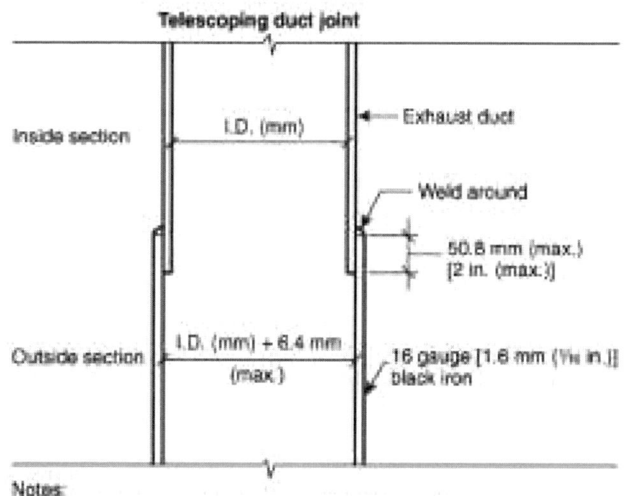

FIGURE 7.5.5.1(a) Telescoping-Type Duct Connection.

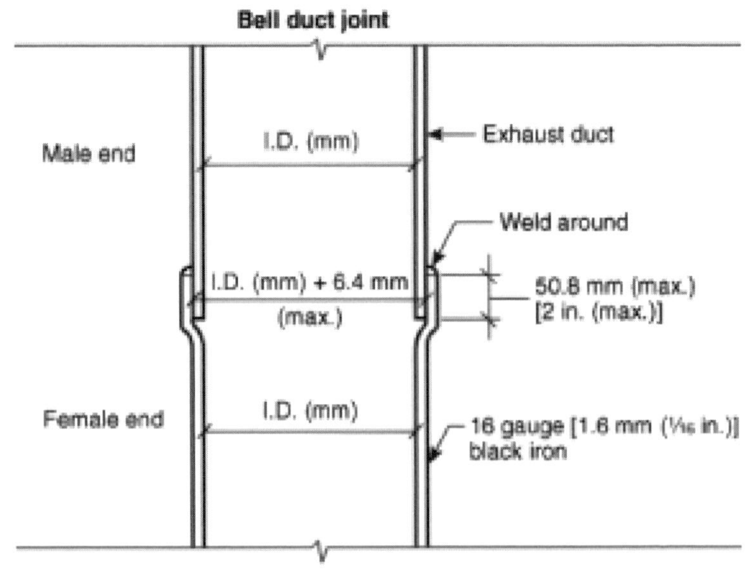

FIGURE 7.5.5.1(b)
Bell-Type Duct Connection.

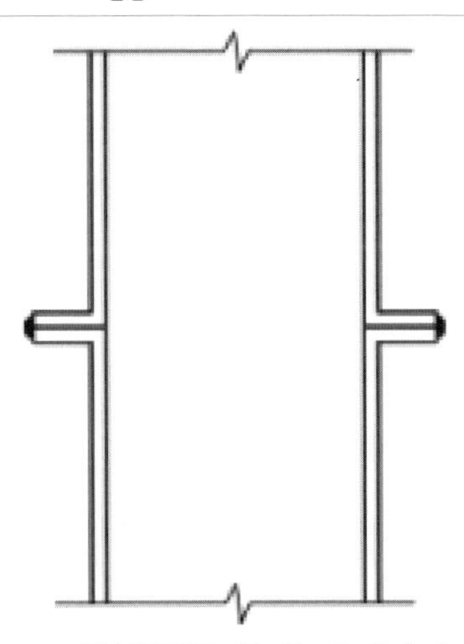

FIGURE 7.5.5.1(c)
Flanged with Edge Weld.

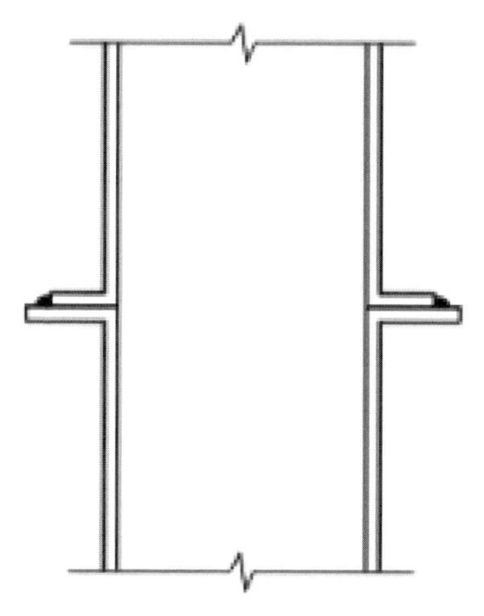

FIGURE 7.5.5.1(d)
Flanged with Filled Weld.

(2) Bell-type joint as shown in Figure 7.5.5.1(b)
(3) Flange with edge weld as shown in Figure 7.5.5.1 (c)
(4) Flange with filled weld as shown in Figure 7.5.5.1 (d)

7.5.5.2 Butt welded connections shall not be permitted.

7.5.5.3 For telescoping and bell-type connections, the inside duct section shall always be uphill of the outside duct section.

7.5.5.4 For telescoping and bell-type connections, the difference between inside dimensions of overlapping sections shall not exceed 6.4 mm (in.)

7.5.5.5 For telescoping and bell-type connections, the overlap shall not exceed 50.8 mm (2 in.)

7.6 Exterior Installations.

7.6.1 The exterior portion of the ductwork shall be vertical wherever possible and shall be installed and supported on the exterior of a building.

7.6.2 Bolts, screws, rivets and other mechanical fasteners shall not penetrate duct walls.

7.6.3 Clearance of any ducts shall comply with Section 4.2

7.6.4 All ducts shall be protected on the exterior by paint or other suitable weather-protective coating.

7.6.5 Ducts constructed of stainless steel shall not be required to have additional paint or weather protective coatings.

7.6.6 Ductwork subject to corrosion shall have minimal contact with the building surface.

7.7 Interior Installations.

7.7.1 Duct Enclosures.

7.7.1.1 In all buildings where vertical fire barriers are penetrated, the ducts shall be enclosed in a continuous enclosure extending from the first penetrated fire barrier and any subsequent fire barriers or concealed spaces, to or through the exterior, so as to maintain the fire resistance rating of the highest fire barrier penetrated.

7.7.1.2 In all buildings more than one story in height and in one-story buildings where the roof-ceiling assembly is required to have a fire resistance rating, the ducts shall be enclosed in a continuous enclosure extending from the lowest fire-rated ceiling or floor above the hood, through any concealed spaces, to or through the roof so as to maintain the integrity of the fire separations required by the applicable building code provisions.

7.7.1.3 The enclosure shall be sealed around the duct at the point of penetration of the first fire-rated barrier after the hood in order to maintain the fire resistance rating of the enclosure.

7.7.1.4 The enclosure shall be vented to the exterior of the building through weather-protected openings.

7.7.1.5 The continuous enclosure provisions shall not be required where a field-applied grease duct enclosure or a factory-built grease duct enclosure (see Section 4.3) is protected with a listed duct-through-penetration protection system equivalent to the fire resistance rating of the assembly being penetrated, and where the materials are installed in accordance with the conditions of the listing and the manufacturer's instructions and are acceptable to the authority having jurisdiction.

7.7.2 Enclosure Fire Resistance Rating and Enclosure Clearance.

7.7.2.1 Fire Resistance Rating.

7.7.2.1.1 Buildings less than four stories in height shall have an enclosure with a fire resistance rating of not less than 1 hour.

7.7.2.1.2 Buildings four stories or more in height shall have an enclosure with a fire resistance rating of not less than 2 hours.

7.7.2.2* Enclosure Clearance.

7.7.2.2.1 Clearance from the duct or the exhaust fan to the interior surface of enclosures of combustible construction shall be not less than 457.2 mm (18 in.)

7.7.2.2.2 Clearance from the duct to the interior surface of enclosures of non-combustible or limited-combustible construction shall be not less than 152.4 mm (6 in.)

7.7.2.2.3 Provisions for reducing clearances as described in Section 4.2 shall not be applicable to enclosures.

7.7.2.2.4 Clearance from the outer surfaces of field-applied grease duct enclosures and factory-built grease duct enclosures to the interior surfaces of construction installed around them shall be permitted to be reduced where the field-applied grease duct enclosure materials and factory-built grease duct enclosures are installed in accordance with the conditions of the listing and manufacturer's instructions and are acceptable to the authority having jurisdiction.

7.7.2.2.5 Field-applied grease duct enclosures and factory built grease duct enclosures shall provide mechanical and structural integrity, resiliency and stability when subjected to expected building environmental conditions, duct movement under general operating conditions and duct movement as a result of interior and exterior fire conditions.

7.7.3 Protection of Coverings and Enclosure Materials.

7.7.3.1 Measures shall be taken to prevent physical damage to any covering or enclosure material.

7.7.3.2 Any damage to the covering or enclosure shall be repaired, and the covering or enclosure shall be restored to meet its intended listing and fire-resistive rating and to be acceptable to the authority having jurisdiction.

7.7.3.3 In the event of a fire within a kitchen exhaust system, the duct, the enclosure, or the covering directly applied to the duct shall be inspected by qualified personnel to determine whether the duct, the enclosure, and the covering directly applied to the duct are structurally sound, capable of maintaining their fire protection functions, suitable for continued operation, and acceptable to the authority having jurisdiction.

7.7.3.4 Listed grease ducts shall be installed in accordance with the terms of the listing and the manufacturer’s instructions.

7.7.4 Enclosure Openings.

7.7.4.1 Where openings in the enclosure walls are provided, they shall be protected by listed fire doors of proper rating.

7.7.4.2 Fire doors shall be installed in accordance with NFPA 80.

7.7.4.3 Openings on other listed materials or products shall be clearly identified and labeled according to the terms of the listing and the manufacturer's instructions and shall be acceptable to the authority having jurisdiction.

7.7.4.4 The fire door shall be readily accessible, aligned, and of sufficient size to allow access to the rated access panels on the ductwork.

7.7.5 Ducts with Enclosure(s).

7.7.5.1 Each duct system shall constitute an individual system serving only exhaust hoods in one fire zone on one floor.

7.7.5.2 Multiple ducts shall not be permitted in a single enclosure unless acceptable to the authority having jurisdiction.

7.8* Termination of Exhaust System.

7.8.1 The exhaust system shall terminate as follows:

(1)*Outside the building with a fan or duct
(2) Through the roof or to the roof from outside, as in 7.8.2, or through a wall, as in 7.8.3

7.8.2 Rooftop Terminations.

7.8.2.1 Rooftop terminations shall be arranged with or provided with the following:

(1) A minimum of 3.05 m (10 ft) of horizontal clearance from the outlet to adjacent buildings, property lines, and air intakes.
(2) A minimum of 1.5 m (5 ft) of horizontal clearance from the outlet (fan housing) to any combustible structure.
(3) A vertical separation of 0.92 m (3 ft) below any exhaust outlets for air intakes within 3.05 m (10 ft) of the exhaust outlet.
(4) The ability to drain grease out of any traps or low points formed in the fan or duct near the termination of the system into a collection container that is non-combustible, closed, rainproof, structurally sound for the service to which it is applied and that will not sustain combustion.
(5) A grease collection device that is applied to exhaust systems that does not inhibit the performance of any fan.

(6) Listed grease collection systems that meet the requirements of 7.8.2.1(4) and 7.8.2.1(5).

(7) A listed grease duct complying with Section 4.4 or ductwork complying with Section 4.5.

(8) A hinged upblast fan supplied with flexible weatherproof electrical cable and service hold-open retainer to permit inspection and cleaning that is listed for commercial cooking equipment with the following conditions:

(a) Where the fan attaches to the ductwork, the ductwork shall be a minimum of 0.46 m (18 in.) away from any roof surface, as shown in Figure 7.8.2.1.

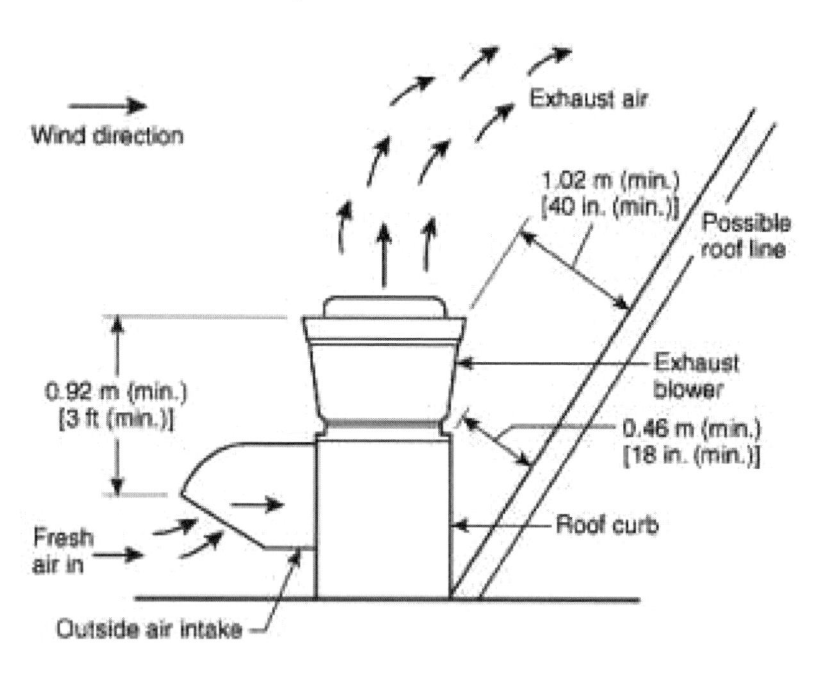

FIGURE 7.8.2.1 Upblast Fan Clearances.

(b) The fan shall discharge a minimum of 1.02 m (40 in.) away from any roof surface, as shown in Figure7.8.2.1.

(9) Other approved fan, provided it meets both of the following:

(a) The fan meets the requirements of 7.8.2.1(3) and 8.1.3.

(b) Its discharge or its extended duct discharge meets the requirements of 7.8.2.1(2). (See 8.1.3.)

7.8.2.2* Fans shall be provided with safe access and a work surface for inspection and cleaning.

7.8.3 Wall Terminations. Wall terminations shall be arranged with or provided with the following properties:

(1) Through a non-combustible wall with a minimum of 3.05 m (10 ft) of clearance from the outlet to adjacent buildings, property lines, grade level, combustible construction, electrical equipment or lines, and the closest point of any air intake or operable door or window at or below the plane of the exhaust termination.

(2) The closest point of any air intake or operable door or window above the plane of the exhaust termination shall be a minimum of 3 m (10 ft) in distance, plus 0.076 m (0.25 ft) for each 1 degree from horizontal, the angle of degree being measured from the center of the exhaust termination to the center of the air intake or operable door or window, as indicated in Figure 7.8.3.

(3) A wall termination in a secured area shall be permitted to be at a lower height above grade if acceptable to the authority having jurisdiction.

(4) The exhaust flow directed perpendicularly outward from the wall face or upward.

(5) All the ductwork pitched to drain the grease back into the hood(s), or with a drain provided to bring the grease back into a container within the building or into a remote grease trap.

(6) A listed grease duct complying with Section 7.4, or other ducts complying with Section 7.5.

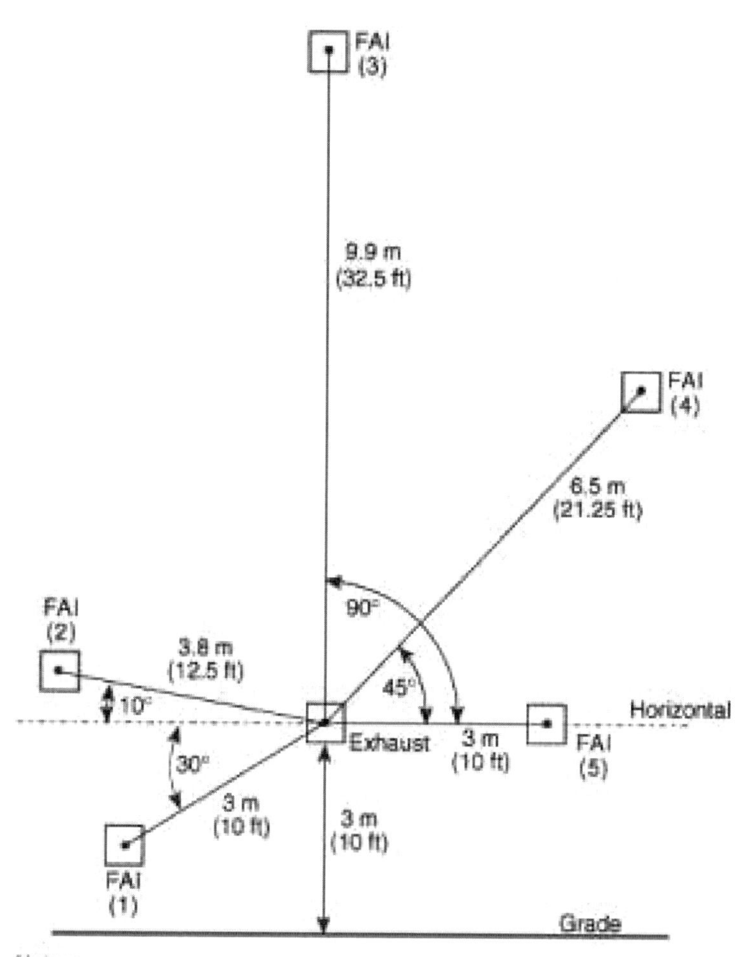

Notes:

1. Fresh air intake (FAI) applies to any air intake, including an operable door or window.
2. Examples:
   (a) FAIs 1 and 5 are on the same plane of exhaust or lower: 3 m (10 ft) min. between closest edges.
   (b) FAIs 2, 3, and 4 are above the plane of exhaust fan: 3 m + .076 m (10 ft + 0.25 ft) per degree between closest edges.

FIGURE 7.8.3 Exhaust Termination Distance from Fresh Air Intake (FAI) or Operable Door or Window.

(7) An approved fan, provided it meets the requirements of 7.8.3(5) and 8.1.1 or 8.1.3.

7.8.4* Rooftop Terminations Through Combustible or Limited-Combustible Walls.

7.8.4.1 Ductwork that exits a building through a combustible or limited-combustible wall to terminate above the roof line shall have wall protection provided in accordance with Section 4.2.

7.8.4.2 Where the ductwork exits the building, the opening shall be sealed and shall include a weather-protected vented opening.

7.8.4.3 Where the ductwork exits through a rated wall, the penetration shall be protected in accordance with 4.4.1.

CHAPTER 8

Air Movement

8.1 Exhaust Fans for Commercial Cooking Equipment

8.1.1* Upblast Exhaust Fans.

8.1.1 Approved upblast fans with motors surrounded by the airstream shall be hinged, supplied with flexible weatherproof electrical cable and service hold-open retainers, and listed for this use.

8.1.1.2 Installation shall conform to the requirements of Section 7.8.

8.1.2* In-Line Exhaust Fans.

8.1.2.1 In-line fans shall be of the type with the motor located outside the airstream and with belts and pulleys protected from the airstream by a greasetight housing.

8.1.2.2 In-line fans shall be connected to the exhaust duct by flanges securely bolted as shown in Figure 8.1.2.2(a) through Figure 8.1.2.2(d) or by a system specifically listed for such use.

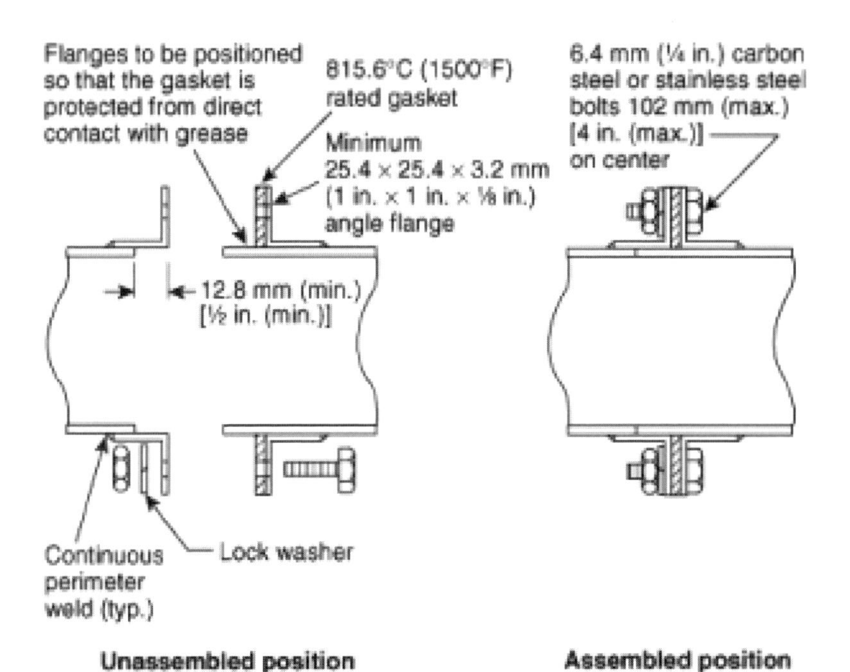

FIGURE 8.1.2.2(a) Typical Section of Duct-to-Fan Connection - Butt Joint Method.

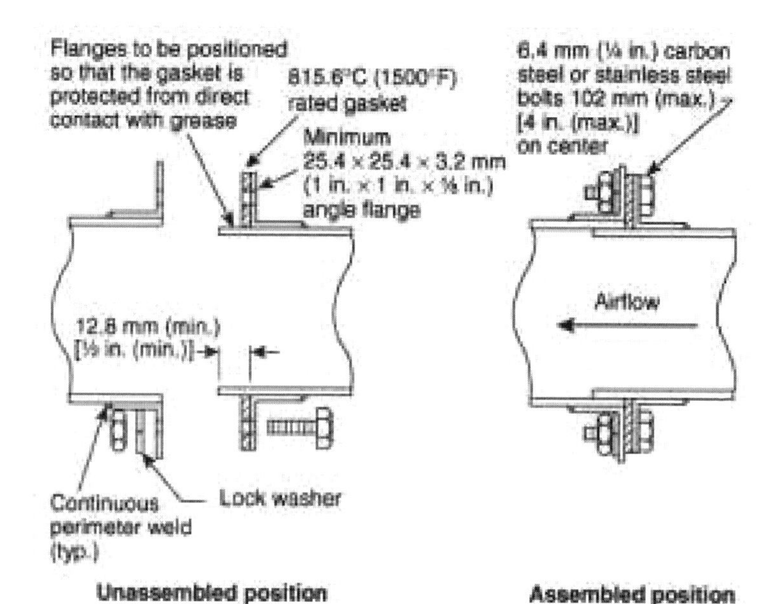

FIGURE 8.1.2.2(b) Typical Section of Duct-to-Fan Connection - Overlapping Method.

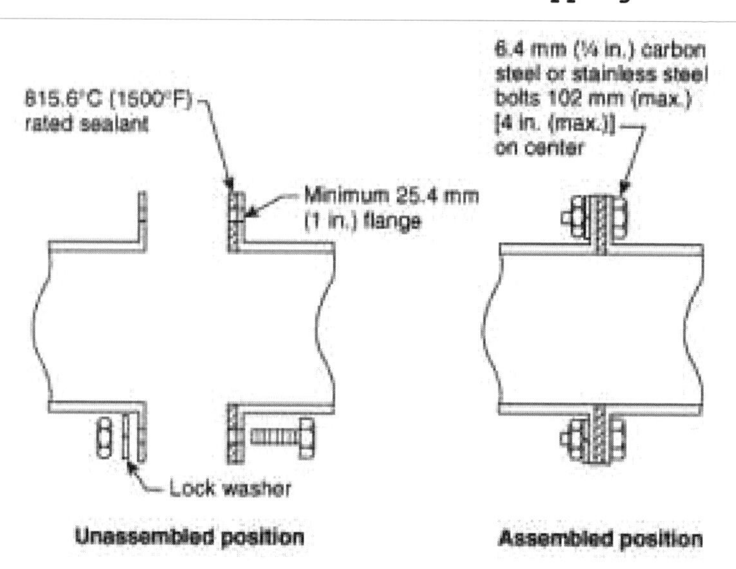

FIGURE 8.1.2.2(c) Typical Section of Duct-to-Fan Connection - Sealant Method.

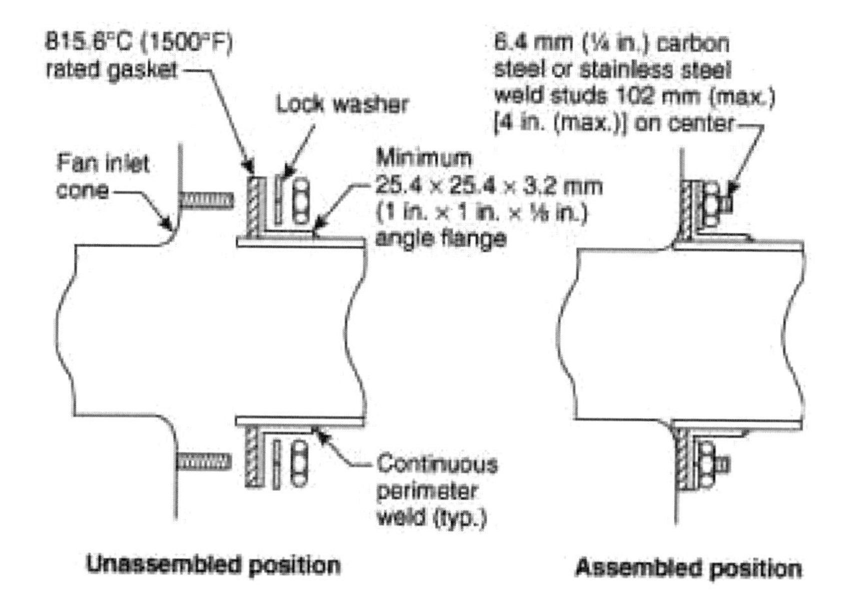

FIGURE 8.1.2.2(d) Typical Section of Duct-to-Fan Connection - Direct to Fan Inlet Cone Method.

8.1.2.3 Flexible connectors shall not be used.

8.1.2.4 If the design or positioning of the fan allows grease to be trapped, a drain directed to a readily accessible and visible grease receptacle not exceeding 3.8 L (1 gal) shall be provided.

8.1.2.5 In-line exhaust fans shall be located in easily accessible areas of adequate size to allow for service or removal.

8.1.2.6 Where the duct system connected to the fan is in an enclosure, the space or room in which the exhaust fan is located shall have the same fire resistance rating as the enclosure.

8.1.3* Utility Set Exhaust Fans.

8.1.3.1 Utility set exhaust fans, if installed at the rooftop termination point, shall meet the requirements of 7.8.2.1(1) through 7.8.2.1(3) and 7.8.2.2.

8.1.3.2 Fans installed within the building shall be located in an accessible area of adequate size to allow for service or removal.

8.1.3.3 Where the duct system connected to the fan is in an enclosure, the space or room in which the exhaust fan is located shall have the same fire resistance rating as the enclosure.

8.1.3.4 The fan shall be connected to the exhaust duct by flanges securely bolted as shown in Figure 8.1.2.2(a) through Figure 8.1.2.2(d) or by a system specifically listed for such use.

8.1.3.5 Flexible connectors shall not be used.

8.1.3.6 Exhaust fans shall have a drain directed to a readily accessible and visible grease receptacle not to exceed 3.8 L (1 gal).

8.1.4 Exhaust Fan Housings. Exhaust fan housings shall be constructed of carbon steel not less than 1.37 mm (0.054 in.) (No. 16 MSG) in thickness or stainless steel not less than 1.09 mm (0.043 in.) (No. 18 MSG) in thickness or, if listed, constructed in accordance with the terms of the listing.

8.1.5 Openings for Cleaning, Servicing, and Inspection.

8.1.5.1 Openings for cleaning, servicing, and inspection shall conform to the requirements of 7.3.7.

8.1.5.2 Clearances shall conform to the requirements of Section 4.2 or, if installed within an enclosure, to the requirements of 7.7.2.2.

8.1.5.3 Upblast fans shall be supplied with an access opening a minimum of 76 mm by 127 mm (3 in. by 5 in.) on the curvature of the outer fan housing to allow for cleaning and inspection of the fan blades.

8.1.6 Wiring and Electrical Equipment. All wiring and electrical equipment shall comply with NFPA 70 (see also Chapter 9).

8.2 Airflow.

8.2.1 Air Velocity.

8.2.1.1* The air velocity through any duct shall be not less than 152.4 m/min (500 ft /min).

8.2.1.2 Transition duct sections that do not exceed 1 m (3 ft) in length and do not contain grease traps shall be permitted to be connected to hoods and exhaust fans that do not meet this velocity.

8.2.2 Air Volume.

8.2.2.1 Exhaust air volumes for hoods shall be of a sufficient level to provide for capture and removal of grease-laden cooking vapors.

8.2.2.2 Test data, performance acceptable to the authority having jurisdiction, or both shall be provided, displayed, or both on request.

8.2.2.3 Lower exhaust air volumes shall be permitted during no-load cooking conditions provided they are sufficient to capture and remove flue gases and residual vapors from cooking equipment.

8.2.3 Exhaust Fan Operation.

8.2.3.1 A hood exhaust fan(s) shall continue to operate after the extinguishing system has been activated unless fan shutdown is required by a listed component of the ventilation system or by the design of the extinguishing system.

8.2.3.2 The hood exhaust fan shall not be required to start upon activation of the extinguishing system if the exhaust fan and all cooking equipment served by the fan have previously been shut down.

8.3* Replacement Air.

8.3.1 Replacement air quantity shall be adequate to prevent negative pressures in the commercial cooking area(s) from exceeding 4.98 kPa (0.02 in. water column).

8.3.2 When its fire-extinguishing system discharges, makeup air supplied internally to a hood shall be shut off.

8.4 Common Duct (Manifold) Systems.

8.4.1* Master kitchen exhaust ducts that serve multiple tenants shall include provision to bleed air from outdoors or from adjacent spaces into the master exhaust duct where required to maintain the necessary minimum air velocity in the master exhaust duct.

8.4.2 Bleed air ducts shall connect to the top or side of the master exhaust duct.

8.4.3 The bleed air duct shall have a fire damper at least 304.8 mm (12 in.) from the exhaust duct connection.

8.4.4 The bleed air duct shall have the same construction and clearance requirements as the main exhaust duct from the connection to the exhaust duct to at least 304.8 mm (12 in.) on both sides of the fire damper.

8.4.5 Each bleed air duct shall have a means of adjusting (e.g., using volume dampers) the bleed air quantity.

8.4.6 Means to adjust the bleed air quantity shall be installed between the fire damper and the source of bleed air.

8.4.7 A bleed air duct shall not be used for the exhaust of grease-laden vapors and shall be so labeled.

8.4.8 Unused tenant exhaust connections to the master exhaust duct that are not used as bleed air connections shall be disconnected and sealed at the main duct.

CHAPTER 9

Auxiliary Equipment

9.1 Dampers.

9.1.1 Dampers shall not be installed in exhaust ducts or exhaust duct systems.

9.1.2 Where specifically listed for such use or where required as part of a listed device or system, dampers in the exhaust ducts or exhaust duct systems shall be permitted.

9.2 Electrical Equipment.

9.2.1 Wiring systems of any type shall not be installed in ducts.

9.2.2 Only where specifically listed for such use shall motors, lights and other electrical devises be permitted to be installed in ducts or hoods or to be located in the path of trav-el of exhaust products.

9.2.3 Lighting Units.

9.2.3.1 Lighting units in hoods shall be list-ed for use over commercial cooking appliances and installed in accordance with the terms of their listing.

9.2.3.2 Lighting units on hoods shall not be located in concealed spaces.

9.2.3.3 Lighting units shall be permitted in concealed spaces where such units are part of a listed exhaust hood.

9.2.3.4 Listed lighting units specifically listed for such use and installed in accordance with the terms of the listing shall be permitted to be installed in concealed spaces.

9.24* All electrical equipment shall be installed in accordance with NFPA 70.

9.3 Other Equipment.

9.3.1 Fume incinerators, thermal recovery units, air pollution control devices, or other devices shall be permitted to be installed in ducts or hoods or to be located in the path of travel of exhaust products where specifically approved for such use.

9.3.2 Downgrading other parts of the exhaust system due to the installation of these approved devices, whether listed or not, shall not be allowed.

9.3.3 Any equipment, listed or otherwise, that provides secondary filtration or air pollution control and that is installed in the path of travel of exhaust products shall be provided with an approved automatic fire-extinguishing system for the protection of the component sections of the equipment and shall include protection of the ductwork downstream of the equipment, whether or not the equipment is provided with a damper.

9.3.4 If the equipment provides a source of ignition, it shall be provided with detection to operate the fire-extinguishing system protecting the equipment.

9.3.5 Where a cooking exhaust system employs an air pollution control device that recirculates air into the building, the requirements of Chapter 13 shall apply.

CHAPTER 10

Fire-Extinguishing Equipment

10.1 General Requirements.

10.1.1 Fire-extinguishing equipment for the protection of grease removal devices, hood exhaust plenums, and exhaust duct systems shall be provided.

10.1.2* Cooking equipment that produces grease-laden vapors and that might be a source of ignition of grease in the hood, grease removal device, or duct shall be protected by fire-extinguishing equipment.

10.2 Types of Equipment.

10.2.1 Fire-extinguishing equipment shall include both automatic fire-extinguishing systems as primary protection and portable fire extinguishers as secondary backup.

COMMENT: The manual control for the suppression system shall be activated before portable extinguishers are used.

10.2.2 A placard identifying the use of the extinguisher as secondary backup means to the automatic fire-extinguishing system shall be conspicuously placed near each portable fire extinguisher intended to be used for protection in the cooking area.

10.2.2.1 The language and wording for the placard shall be approved by the authority having jurisdiction.

COMMENT: The language and wording is being left up to the discretion of the AHJ due to the various makeup of restaurant employee's diverse ethnicity throughout the country. For example, it does not make sense to place a placard in English if the kitchen staff only reads Spanish or Chinese.

10.2.3* Automatic fire-extinguishing systems shall comply with UL 300 or other equivalent standards and shall be installed in accordance with the requirements of the listing.

10.2.3.1 In existing systems, when changes in the cooking media, positioning, or replacement of cooking equipment occur, the fire-extinguishing system shall be made to comply with 10.2.3.

10.2.4 Grease removal devices, hood exhaust plenums, exhaust ducts, and cooking equipment that are not addressed in UL 300 or other equivalent test standards shall be protected with an automatic fire-extinguishing system(s) in accordance with the applicable NFPA standard(s) and all local building and fire codes and shall be approved by the authority having jurisdiction.

10.2.5 Automatic fire-extinguishing equipment provided as part of listed recirculating systems shall comply with UL 197.

10.2.6 Automatic fire-extinguishing systems shall be installed in accordance with the terms of their listing, the manufacturer's instructions, and the following standards where applicable.

(1) NFPA 12
(2) NFPA 13
(3) NFPA 17
(4) NFPA 17A

10.2.7 Modifications to Existing Hood Systems.

10.2.7.1 Any abandoned pipe or conduit from a previous installation shall be removed from within the hood, plenum, and exhaust duct.

10.2.7.2 Penetrations and holes resulting from the removal of conduit or piping shall be sealed with listed or equivalent liquid tight sealing devices.

COMMENT: The removal of a dry chemical system being replaced with a UL 300 system causes penetration holes to remain after the removal of the piping. This is another concern that should be identified during the post fire examination of the ventilation system.

10.2.7.3 The addition of obstructions to spray patterns from the cooking appliance nozzle(s) such as baffle plates, shelves, or any modification shall not be permitted.

10.2.7.4 Changes or modifications to the hazard after installation of the fire-extinguishing systems shall result in re-evaluation of the system design by a properly trained and qualified person(s) or company.

10.2.8 Fixed Baffle Hoods with Water Wash.

10.2.8.1 Grease removal devices, hood exhaust plenums, and exhaust ducts requiring protection in accordance with 10.1.1 shall be permitted to be protected by a listed fixed baffle hood containing a constant or fire-actuated water-wash system that is listed and in compliance with UL 300 or other equivalent standards and shall be installed in accordance with the requirements of their listing.

10.2.8.2 Each such area not provided with a listed water-wash extinguishing system shall be provided with a fire-extinguishing system listed for the purpose.

COMMENT: There is no dry or wet chemical suppression system that has been listed without duct and plenum protection.

10.2.8.3 The water for listed fixed baffle hood assemblies shall be permitted to be supplied from the domestic water supply when the minimum water pressure and flow are provided in accordance with the terms of the listing.

10.2.8.4 The water supply shall be controlled by a supervised water supply control valve.

10.2.8.5 The water wash in a fixed baffle hood, specifically listed to extinguish a fire, shall be activated by the cooking equipment extinguishing system.

10.2.8.6 A water-wash system approved to be used for protection of the grease removal device(s), hood exhaust plenum(s), exhaust duct(s), or combination thereof shall include instruction and appropriate electrical interface for simultaneous activation of the water-wash system from an automatic fire-extinguishing system, where the automatic fire-extinguishing system is used for cooking equipment protection only.

10.2.8.7 Where the fire-extinguishing system provides protection for the cooking equipment, hood, and duct, activation of the water wash shall not be required.

10.2.9 The water required for listed automatic fire-extinguishing systems shall be permitted to be supplied from the domestic water supply where the minimum water pressure and flow are provided in accordance with the terms of the listing. The water supply shall be controlled by a supervised water supply control valve.

10.2.10 Water Valve Supervision. Valves controlling the water supply to listed fixed baffle hood assemblies, automatic fire-extinguishing systems, or both shall be listed indicating type of valve and shall be supervised open by one of the following methods:

(1) Central station, proprietary, or remote station alarm service

(2) Local alarm service that will cause the sounding of an audible signal at a constantly attended point

(3) Locking valves open

(4) *Sealing of valves and approved weekly recorded inspection.

10.3 Simultaneous Operation.

10.3.1 Fixed pipe extinguishing systems in a single hazard area (see Section 3.3 for the definition of Single Hazard Area) shall be arranged for simultaneous automatic operation upon actuation of any one of the systems.

10.3.2 Simultaneous operation shall not be required where the fixed pipe extinguishing system is an automatic sprinkler system.

10.3.3 Simultaneous operation shall not be required where dry or wet chemical system shall be permitted to be used to protect common exhaust ductwork by one of the methods specified in NFPA 17 or NFPA 17A.

10.4 Fuel Shutoff.

10.4.1 Upon activation of any fire-extinguishing system for a cooking operation, all sources of fuel and electric power that produce heat to all equipment requiring protection by that system shall automatically shut off.

COMMENT: This allows a cooling process of oil to occur and therefore the breakdown and ignition of the saponicated oil will not occur.

10.4.2 Steam supplied from an external source shall not be required to automatically shut off.

10.4.3 Any gas appliance not requiring protection, but located under the same ventilating equipment, shall also automatically shut off upon activation of any extinguishing system.

10.4.4 Shutoff devices shall require manual reset.

10.5 Manual Activation.

10.5.1 A readily accessible means for manual activation shall be located between 1067 mm and 1219 mm (42 in. and 48 in.) above the floor, be accessible in the event of a fire, be located in a path of egress, and clearly identify the hazard protected.

10.5.2 The automatic and manual means of system activation external to the control head or releasing device shall be separate and independent of each other so that failure of one will not impair the operation of the other.

10.5.3 The manual means of system activation shall be permitted to be common with the automatic means if the manual activation device is located between the control head or releasing device and the first fusible link.

COMMENT: Pre UL 300 wet chemical and dry chemical system manufacturer's allowed the manual pull station to be installed at the end of the automatic fuse link line. The problems is that if a failure of the automatic line occurs then the manual pull station at the end of line would probably be effected and failure of manual operation will occur.

10.5.4 An automatic sprinkler system shall not require a manual means of system activation.

10.5.5 The means for manual activation shall be mechanical or rely on electrical power for activation in accordance with 10.5.6.

10.5.6 Electrical power shall be permitted to be used for manual activation if a standby power supply is provided or if supervision is provided in accordance with Section 10.7.

10.6 System Annunciation.

10.6.1 Upon activation of an automatic fire-extinguishing system, an audible alarm or visual indicator shall be provided to show that the system has activated.

10.6.2 Where a fire alarm signaling system is serving the occupancy where the extinguishing system is located, the activation of the automatic fire-extinguishing system shall activate the fire alarm signaling system.

10.7 System Supervision.

10.7.1 Where electrical power is required to operate the automatic fire-extinguishing system, it shall be monitored by a supervisory alarm, with a standby power supply provided.

10.7.2 System supervision shall not be required where an automatic fire-extinguishing system(s) includes automatic mechanical detection and actuation as a backup detection system.

10.7.3 System supervision shall not be required where a fire-extinguishing system(s) is interconnected or interlocked with the cooking equipment power source(s) so that if the fire-extinguishing system becomes inoperable due to power failure, all sources of fuel or electric power that produce heat to all cooking equipment serviced by that hood shall automatically shut off.

10.7.4 System supervision shall not be required where an automatic fire-extinguishing system, including automatic mechanical detection and actuation, is electrically connected to a listed fire-actuated water-wash system for simultaneous operation of both systems.

10.8 Special Design and Application.

10.8.1 Hoods containing automatic fire-extinguishing systems are protected areas; therefore, these hoods are not considered obstructions to overhead sprinkler systems and shall not require floor coverage underneath.

10.8.2 A single listed detection device shall be permitted for more than one appliance when installed in accordance with the terms of the listing.

10.9 Review and Certification.

10.9.1 Where required, complete drawings of the system installation, including the hood(s), exhaust duct(s), and appliances, along with the interface of the fire-extinguishing system detectors, piping, nozzles, fuel shutoff devices, agent storage container(s), and manual actuation device(s), shall be submitted to the authority having jurisdiction.

10.9.2* Installation Requirements.

10.9.2.1 Installation of systems shall be performed only by persons properly trained and qualified to install the specific system being provided.

10.9.2.2 The installer shall provide certification to the authority having jurisdiction that the installation is in agreement with the terms of the listing and the manufacturer's instructions and/or approved design.

10.10 Portable Fire Extinguishers.

10.10.1* Portable fire extinguishers shall be installed in kitchen cooking areas in accordance with NFPA 10 and shall be specifically listed for such use.

10.10.2 Extinguishers shall use agents that saponify upon contact with hot grease such as sodium bicarbonate and potassium bicarbonate dry chemical and potassium carbonate solutions.

10.10.3 Class B gas-type portables shall not be permitted in kitchen cooking areas.

10.10.4 The manufacturer's recommendations shall be followed.

10.10.5 Other fire extinguishers in the kitchen area shall be installed in accordance with NFPA 10.

CHAPTER 11

Procedures for the Use and Maintenance of Equipment

11.1 Operating Procedures

11.1.1 Exhaust systems shall be operated whenever cooking equipment is turned on.

11.1.2 Filter-equipped exhaust systems shall not be operated with filters removed.

11.1.3 Openings provided for replacing air exhausted through ventilating equipment shall not be restricted by covers, dampers, or any other means that would reduce the operating efficiency of the exhaust system.

11.1.4 Instructions for manually operating the

fire-extinguishing system shall be posted conspicuously in the kitchen and shall be reviewed with employees by the management.

11.1.5 Listed exhaust hoods shall be operated in accordance with the terms of their listings and the manufacturer's instructions.

11.1.6 Cooking equipment shall not be operated while its fire-extinguishing system or exhaust system is non-operational or otherwise impaired.

11.1.7 Secondary filtration and pollution control equipment shall be operated in accordance with the terms of their listing and the manufacturer's recommendations.

11.1.8 Inspection and maintenance of equipment allowed in 9.3.1 shall be conducted by properly trained and qualified persons at a frequency determined by the manufacturer's instructions or equipment listing.

11.2 Inspection of Fire-Extinguishing Systems.

11.2.1* Maintenance of the fire-extinguishing systems and listed exhaust hoods containing a constant or fire activated water system that is listed to extinguish a fire in the grease removal devices, hood exhaust plenums, and exhaust ducts shall be made by properly trained, qualified and certified person(s) or company acceptable to the authority having jurisdiction at least every 6 months.

11.2.2 All actuation components, including remote manual pull stations, mechanical or electrical devices, detectors, actuators and fire-actuated dampers shall be checked for proper operation during the inspection in accordance with the manufacturer's listed procedures.

11.2.3 In addition to these requirements, the specific inspection requirements of the applicable NFPA standard shall also be followed.

11.2.4* Fusible links (including fusible links on fire damper assemblies) and automatic sprinkler heads shall be replaced at least semi-annually or more frequently if necessary where required by the manufacturer.

11.2.5 The year of manufacture and the date of installation of the fusible links shall be marked on the system inspection tag. The tag shall be signed or initialed by the installer.

11.2.6 Other detection devices not including fusible links and automatic sprinklers shall be serviced or replaced in accordance with the manufacturer's recommendations.

11.2.7 Where automatic bulb-type sprinklers or spray nozzles are used and annual examination shows no buildup of grease or other material on the sprinkler or spray nozzles, annual replacement shall not be required.

11.2.8 Where required, certificates of inspection and maintenance shall be forwarded to the authority having jurisdiction.

11.3* Inspection for Grease Buildup. The entire exhaust system shall be inspected for grease buildup by a properly trained, qualified and certified company or person(s) acceptable to the authority having jurisdiction in accordance with Table 11.3.

Table 11.3
Schedule of Inspection for Grease Buildup

Type or Volume of Cooking Frequency	Frequency
Systems serving solid fuel cooking operations	Monthly
Systems serving high-volume cooking operations, such as 24-hour cooking, charbroiling, or wok cooking	Quarterly
Systems serving moderate-volume cooking operations	Semi-annually
Systems serving low-volume cooking operations, such as churches, day camps, seasonal businesses, or senior centers	Annually

11.4 Cleaning of Exhaust Systems.

11.4.1 Upon inspection, if found to be contaminated with deposits from grease-laden vapors, the contaminated portions of the exhaust system shall be cleaned by a properly trained, qualified and certified company or person(s) acceptable to the authority having jurisdiction.

11.4.2* Hoods, grease removal devices, fans, ducts and other appurtenances shall be cleaned to remove combustible contaminants prior to surfaces becoming heavily contaminated with grease or oily sludge.

11.4.3 At the start of the cleaning process, electrical switches that could be activated accidentally shall be locked out.

11.4.4 Components of the fire suppression system shall not be rendered inoperable during the cleaning process.

11.4.5 Fire-extinguishing systems shall be rendered inoperable during the cleaning process where serviced by properly trained and qualified persons.

11.4.6 Flammable solvents or other flammable cleaning aids shall not be used.

11.4.7 Cleaning chemicals shall not be applied on fusible links or other detection devices of the automatic extinguishing system.

11.4.8 After the exhaust system is cleaned, it shall not be coated with powder or other substance.

11.4.9 When cleaning procedures are completed, all access panels (doors) and cover plates shall be restored to their normal operational condition.

11.4.10 Dampers and diffusers shall be positioned for proper airflow.

11.4.11 When cleaning procedures are completed, all electrical switches and system components shall be returned to an operable state.

11.4.12 When a vent cleaning service is used, a certificate showing date of inspection or cleaning shall be maintained on the premises.

11.4.13 After cleaning is completed, the vent cleaning contractor shall place on display within the kitchen area a label indicating the date cleaned, the name of the servicing company, and areas not cleaned.

11.4.14 Where required, certificates of inspection and cleaning shall be submitted to the authority having jurisdiction.

11.5 Cooking Equipment Maintenance.

11.5.1 An inspection and servicing of the cooking equipment shall be made at least annually by properly trained and qualified persons.

11.5.2 Cooking equipment that collects grease below the surface, behind the equipment, or in cooking equipment flue gas exhaust, such as griddles or charbroilers, shall be inspected and, if found with grease accumulation, cleaned by a properly trained, qualified and certified person acceptable to the authority having jurisdiction.

CHAPTER 12

Minimum Safety Requirements for Cooking Equipment

12.1 Cooking Equipment.

12.1.1 Cooking equipment shall be approved based on one of the following criteria:

(1) Listings by a testing laboratory
(2) Test data acceptable to the authority having jurisdiction

12.1.2 Installation.

12.1.2.1 All listed appliances shall be installed in accordance with the terms of their listings and the manufacturer's instructions.

12.1.2.2* Cooking appliances requiring protection shall not be moved, modified, or rearranged without prior re-evaluation of the fire-extinguishing system by the system installer or servicing agent, unless otherwise allowed by the design of the fire-extinguishing system.

12.1.2.3 The fire-extinguishing system shall not require re-evaluation where the cooking appliances are moved to perform maintenance and cleaning provided the appliances are returned to approve design location prior to cooking operations, and any disconnected fire-extinguishing system nozzles attached to the appliances are reconnected in accordance with the manufacturer's listed design manual.

12.1.2.3.1 An approved method shall be provided that will ensure that the appliance is returned to an approved design location.

12.1.24 All deep-fat fryers shall be installed with at least a 406 mm (16 in.) space between the fryer and surface flames from adjacent cooking equipment.

12.1.2.5 Where a steel or tempered glass baffle plate is installed at a minimum 203 mm (8 in.) in height between the fryer and surface flames of the adjacent appliance, the requirement for a 406 mm (16 in.) space shall not apply.

12.1.2.5.1 If the fryer and the surface flames are at different horizontal planes, the minimum height of 203 mm (8 in.) shall be measured from the higher of the two.

12.2 Operating Controls. Deep-fat fryers shall be equipped with a separate high-limit control in addition to the adjustable operating control (thermostat) to shut off fuel or energy when the fat temperature reaches 246EC (47EF) at 25.4 mm (1 in.) below the surface.

CHAPTER 13
Recirculating System

13.1 General Requirements. Recirculating systems containing or for use with appliances used in processes producing smoke or grease-laden vapors shall be equipped with components complying with the following:

(1) The clearance requirements of Section 4.2.

(2) A hood complying with the requirements of Chapter 5.

(3) Grease removal devices complying with Chapter 6.

(4) The air movement requirements of 8.2.1.2 and 8.2.2.3.

(5) Auxiliary equipment (such as particulate and odor removal devices) complying with Chapter 9.

(6) Fire-extinguishing equipment complying with the requirements of Chapter 10 with the exception of 10.1.1 and 10.5.1, which shall not apply.

(7) The use and maintenance requirements of Chapter 11.

(8) The minimum safety requirements of Chapter 12.

(9) All the requirements of Chapter 13.

13.2 Design Restrictions. All recirculating systems shall comply with the requirements of Section 13.2.

13.2.1 Only gas-fueled or electrically fueled cooking appliances shall be used.

13.2.2 Listed gas fueled equipment deigned for use with specific recirculating systems shall have the flue outlets connected in the intended manner.

13.2.3 Gas-fueled appliances shall have a minimum 457.2 mm (18 in.) clearance from the flue outlet to the filter inlet in accordance with 6.2.2 and shall meet the installation requirements of NFPA 54 or NFPA 58.

13.2.4 Recirculating systems shall be listed with a testing laboratory.

13.2.5 There shall be no substitution or exchange of cooking appliances, filter components, blower components or fire-extinguishing system components that would violate the listing of the appliance.

13.2.6 A recirculating system shall not use cooking equipment that exceeds the recirculating system's labeled maximum limits for that type of equipment, stated in maximum energy input, maximum cooking temperature, and maximum square area of cooking surface or cubic volume of cooking cavity.

13.2.7 The listing label shall show the type(s) of cooking equipment tested and the maximum limits specified in 13.2.6.

13.2.8 A fire-actuated damper shall be installed at the exit outlet of the system.

13.2.9 The fire damper shall be constructed of at least the same gauge as the shell.

13.2.10 The actuation device for the fire damper shall have a maximum temperature rating of 190EC (375EF).

13.2.11 The power supply of any electrostatic precipitator (ESP) shall be of the "cold spark," ferro-resonant type in which the voltage falls off as the current draw of a short increases.

13.2.12 Listing evaluation shall include the following:

(1) Capture and containment of vapors at published and labeled airflows
(2) Grease discharge at the exhaust outlet of the system not to exceed an average of 5 mg/m3 of exhausted air sampled from that equipment at maximum amount of product that is capable of being processed over a continuous 8-hour test per EPA Test Method 202 with the system operating at its minimum listed airflow
(3) Listing and labeling of clearance to combustibles from all sides, top and bottom
(4) Electrical connection in the field in accordance with NFPA 70
(5) Interlocks on all removable components that lie in the path of airflow within the unit to ensure that they are in place during operation of the cooking appliance.

13.3 Interlocks.

13.3.1 The recirculating system shall be provided with interlocks of all critical components and operations as indicated in 13.3.2 through 13.3.4 such that, if any of these interlocks are interrupted, the cooking appliances shall not be able to operate.

13.3.2 All closure panels encompassing airflow sections shall have interlocks to ensure the panels are in place and fully sealed.

13.3.3 Each filter component (grease and odor) shall have an interlock to prove the component is in place.

13.3.4 ESP Interlocks.

13.3.4.1 Each ESP shall have a sensor to prove its performance is as designed, with no interruption of the power to exceed 2 minutes.

13.3.4.2 The sensor shall be a manual reset device or circuit.

13.3.5 Airflow Switch or Transducer.

13.3.5.1 An airflow switch or transducer shall be provided after the last filer component to ensure that a minimum airflow is maintained.

13.3.5.2 The airflow switch or transducer shall open the interlock circuit when the airflow falls 25 percent below the system's normal operating flow or 10 percent below its listed minimum rating, whichever is lower.

13.3.5.3 The airflow switch or transducer shall be a manual reset device or circuit.

13.4 Location and Application Restrictions.

13.4.1 The location of recirculating systems shall be approved by the authority having jurisdiction.

13.4.2 Items to be reviewed in the fire risk assessment shall include, but not be limited to, life safety, combustibility of surroundings, proximity to air vents, and total fuel load.

13.5.1 In addition to the appliance nozzle(s), a recirculating system shall be listed with the appropriate fire protection for grease filters, grease filtration, odor filtration units, and ductwork, where applicable.

13.5.2 In addition to any other fire-extinguishing activation device, there shall be a fire-extinguishing system activation device installed downstream of any ESP.

13.5.3 The requirements of Section 10.6 shall also apply to recirculating system locations.

13.5.4 A means of manual activation of the fire-extinguishing system shall be provided in an area where it is safely accessible in the event of a fire in the appliance.

13.5.5 The manual activation device for the fire-extinguishing system shall be clearly identified.

13.6 Use and Maintenance.

13.6.1 Automatic or manual covers on cooking appliances, especially fryers, shall not interfere with the application of the fire suppression system.

13.6.2 All filters shall be cleaned or replaced in accordance with the manufacturer's instructions.

13.6.3 All ESPs shall be cleaned a minimum of once per week following the manufacturer's cleaning instructions.

13.6.4 The entire hood plenum and the blower section shall be cleaned a minimum of once every 3 months.

13.6.5 Inspection and testing of the total operation and all safety interlocks in accordance with the manufacturer's instructions shall be performed by qualified service personnel a minimum of once every 6 months or more frequently if required.

13.6.6 Fire-extinguishing equipment shall be inspected in accordance with Section 11.2.

13.6.7 A signed and dated log of maintenance as performed in accordance with 13.6.4 and 13.6.5 shall be available on the premises for use by the authority having jurisdiction.

Chapter 14

Solid Fuel Cooking Operations

14.1 Venting Application. Venting requirements of solid fuel cooking operations shall be determined in accordance with 14.1.1 through 14.1.7.

14.1.1 Where solid fuel cooking equipment is required by the manufacturer to have a natural draft, the vent shall comply with Section 14.4.

14.1.2 Where the solid fuel cooking equipment has a self-contained top, is the only appliance to be vented in an isolated space (except for a single water heater with its own separate vent), has a separate makeup air system, and is provided with supply and return air (not supplied or returned from other spaces), the system shall comply with Sections 14.4 and 14.6.

14.1.3 Where the solid fuel cooking equipment is located in a space with other vented equipment, all vented equipment shall have an exhaust system interlocked with a makeup air system for the space per Section **14.6.**

14.1.4 Natural draft ventilation systems and power-exhausted ventilation systems shall comply with Sections 14.3, 14.4, and 14.6.

14.1.5 Where a solid fuel cooking appliance allows effluent to escape from the appliance opening, this opening shall be covered by a hood and an exhaust system that meets the requirements of Sections 14.3, 14.4, and 14.6.

14.1.6 Solid fuel cooking operations shall have spark arresters to minimize the passage of airborne sparks and embers into plenums and ducts.

14.1.7 Where the solid fuel cooking operation is not located under a hood, a spark arrester shall be provided to minimize the passage of sparks and embers into flues and chimneys.

14.2 Location of Appliances.

14.2.1 Every appliance shall be located with respect to building construction and other equipment so as to permit access to the appliance.

14.2.2* Solid fuel cooking appliances shall not be installed in confined spaces.

14.2.3 Solid fuel cooking appliances listed for installation in confined spaces such as alcoves shall be installed in accordance with the terms of the listing and the manufacturer's instructions.

14.2.4 Solid fuel cooking appliances shall not be installed in any location where gasoline or any other flammable vapors or gases are present.

14.3 Hoods for Solid Fuel Cooking.

14.3.1 Hoods shall be sized and located in a manner capable of capturing and containing all of the effluent discharging from the appliances.

14.3.2 The hood and its exhaust system shall comply with the requirements of Chapters 5 through 10.

14.3.3 All solid fuel cooking equipment served by hoods and duct systems shall be separate from all other exhaust systems.

14.3.4 Cooking equipment not requiring automatic fire-extinguishing equipment (per the provisions of Chapter 10) shall be permitted to be installed under a common hood with solid fuel cooking equipment that is served by a duct system separate from all other exhaust systems.

14.4 Exhaust for Solid Fuel Cooking. Where a hood is not required, in buildings where the duct system is three stories or less in height, a duct complying with Chapter 7 shall be provided.

14.4.1 If a hood is used in buildings where the duct system is three stories or less in height, the duct system shall comply with Chapter 7.

14.4.2 A listed or approved grease duct system that is four stories in height or greater shall be provided for solid fuel cooking exhaust systems.

14.4.3 Where a hood is used, the duct system shall conform with the requirements of Chapter 7.

14.4.4 Wall terminations of solid fuel exhaust systems shall be prohibited.

14.5 Grease Removal Devices for Solid Fuel Cooking.

14.5.1 Grease removal devices shall be constructed of steel or stainless steel or be approved for solid fuel cooking.

14.5.2 If airborne sparks and embers can be generated by the solid fuel cooking operations, spark arrester devices shall be used prior to the grease removal device to minimize the entrance of these sparks and embers into the grease removal device and into the hood and the duct system.

14.5.3 Filters shall be a minimum of 1.2m (4 ft) above the appliance cooking surface.

14.6 Air Movement for Solid Fuel Cooking.

14.6.1 Exhaust system requirements shall comply with Chapter 8 for hooded operation or shall be installed in accordance with the manufacturers recommendations for unhooded applications.

14.6.2 A replacement or makeup air system shall be provided to ensure a positive supply of replacement air at all times during cooking operations.

14.6.3 Makeup air systems serving solid fuel cooking operations shall be interlocked with the exhaust air system and powered, if necessary, to prevent the space from attaining a negative pressure while the solid fuel appliance is in operation.

14.7 Fire-Extinguishing Equipment for Solid Fuel Cooking.

14.7.1 Solid fuel cooking appliances that produce grease laden vapors shall be protected by listed fire-extinguishing equipment.

14.7.2 Where acceptable to the authority having jurisdiction, solid fuel-burning cooking appliances constructed of solid masonry or reinforced portland or refractory cement concrete and vented in accordance with NFPA 211 shall not require fixed automatic fire-extinguishing equipment.

14.7.3 Listed fire-extinguishing equipment shall be provided for the protection of grease removal devices, hoods, and duct systems.

14.7.4 Where acceptable to the authority having jurisdiction, solid fuel-burning cooking appliances constructed of solid masonry or reinforced portland or refractory cement concrete and vented in accordance with NFPA 211 shall not require automatic fire-extinguishing equipment for the protection of grease removal devices, hoods, and duct systems.

14.7.5 Listed fire-extinguishing equipment for solid fuel-burning cooking appliances, where required, shall comply with Chapter 10 and shall use water-based agents.

14.7.6 Fire-extinguishing equipment shall be rated and designed to extinguish solid fuel cooking fires, in accordance with the manufacturer's recommendations.

14.7.7 The fire-extinguishing equipment shall be of sufficient size to totally extinguish fire in the entire hazard area and prevent reignition of the fuel.

14.7.8 All solid fuel appliances (whether or not under a hood) with fire boxes of 0.14 m^3 (5 ft^3) volume or less shall have at least a listed 2-A rated water-type fire extinguisher or a 6 L (1.6 gal) wet chemical fire extinguisher listed for Class K fires in accordance with NFPA 10 in the immediate vicinity of the appliance.

14.7.9 Hose Protection.

14.7.9.1 Solid fuel appliances with fireboxes exceeding 0.14 m^3 (5 ft^3) shall be provided with a fixed water pipe system with a hose in the kitchen capable of reaching the firebox.

14.7.9.1.1 The hose shall be equipped with an adjustable nozzle capable of producing a fine to medium spray or mist.

14.7.9.1.2 The nozzle shall be of the type that cannot produce a straight stream.

14.7.9.2 The system shall have a minimum operating pressure of 275.8 kPa (40 psi) and shall provide a minimum of 19 L/min (5 gpm).

14.7.10 Fire suppression for fuel storage areas shall comply with Section 14.9 of this standard.

14.7.11 In addition to the requirements of 14.7.8 through 14.7.10, where any solid fuel cooking appliance is also provided with auxiliary electric, gas, oil, or other fuel for ignition or supplemental heat and the appliance is also served by any portion of a fire-extinguishing system complying with Chapter 10, such auxiliary fuel shall be shut off on actuation of the fire-extinguishing system.

14.8 Procedures for Inspection, Cleaning, and Maintenance for Solid Fuel Cooking. Solid fuel cooking appliances shall be inspected, cleaned, and maintained in accordance with the procedures outlined in Chapter 11 and with 14.8.1 through 14.8.5.

14.8.1 The combustion chamber shall be scraped clean to its original surface once each week and shall be inspected for deterioration or defects.

14.8.2 Any significant deterioration or defect that might weaken the chamber or reduce its insulation capability shall be immediately repaired.

14.8.3 The flue or chimney shall be inspected weekly for the following conditions:

(1) Residue that might begin to restrict the vent or create an additional fuel source

(2) Corrosion or physical damage that might reduce the flue's capability to contain the effluent

14.8.3.1 The flue or chimney shall be cleaned before these conditions exist.

14.8.3.2 The flue or chimney shall be repaired

or replaced if any unsafe condition is evident.

14.8.4 Spark arrester screens located at the entrance of the flue or in the hood assembly shall be cleaned prior to their becoming heavily contaminated and restricted.

14.8.5 Filters and filtration devices installed in a hood shall be cleaned per 14.8.4.

14.9 Minimum Safety Requirements: Fuel Storage, Handling, and Ash Removal for Solid Fuel Cooking.

14.9.1 Installation Clearances.

14.9.1.1 Solid fuel cooking appliances shall be installed on floors of non-combustible construction that extend 0.92 m (3 ft) in all directions from the appliance.

14.9.1.2 Floors with non-combustible surfaces shall be permitted to be used where they have been approved for such use by the authority having jurisdiction.

14.9.1.3 Floor assemblies that have been listed for solid fuel appliance applications shall be permitted to be used.

14.9.1.4 Solid fuel cooking appliances that have been listed for zero clearance to combustibles on the bottom and sides and have an approved hearth extending 0.92 m (3 ft) in all directions from the service door(s) shall be permitted to be used on combustible floors.

14.9.1.5 Combustible and limited-combustible surfaces or construction within 0.92 m (3 ft) of the sides or 1.8 m (6 ft) above a solid fuel cooking appliance shall be protected in a manner acceptable to the authority having jurisdiction.

14.9.1.6 Solid fuel cooking appliances that are specifically listed for less clearance to combustibles shall be permitted to be installed in accordance with the requirements of the listing and the manufacturer's instructions.

14.9.2 Solid Fuel Storage.

14.9.2.1 Fuel storage shall not exceed a one-day supply where storage is in the same room as the solid fuel appliance or in the same room as the fuel-loading or clean-out doors.

14.9.2.2 Fuel shall not be stored above any heat-producing appliance or vent or closer than 0.92 m (3 ft) to any portion of a solid fuel appliance constructed of metal or to any other cooking appliance that could ignite the fuel.

14.9.2.3 Fuel shall be permitted to be stored closer than the requirements of 14.9.2.2 where a solid fuel appliance or other cooking appliance is listed or approved for less clearance to combustibles.

14.9.2.4 Fuel shall not be stored in the path of the ash removal.

14.9.2.5 Where stored in the same building as the solid fuel appliance, fuel shall be stored only in an area with walls, floor, and ceiling of non-combustible construction extending at least 0.92 m (3 ft) past the outside dimensions of the storage pile.

14.9.2.6 Fuel shall be permitted in the protected areas where combustible or limited-combustible materials are protected in accordance with 4.2.3.

14.9.2.7 Fuel shall be separated from all flammable liquids, all ignition sources, all chemicals, and all food supplies and packaging goods.

14.9.2.8 All fuel storage areas shall be provided with a sprinkler system meeting the requirements of NFPA 13 and acceptable to the authority having jurisdiction.

14.9.2.9 Where acceptable to the authority having jurisdiction, fuel storage areas shall be permitted to be protected with a fixed pipe system capable of reaching all parts of the area and meeting the requirements of 14.7.1 through 14.7.11.

14.9.2.10 The portable fire extinguisher specified in 14.7.8 shall be permitted to be used for a solid fuel pile, provided that the fuel pile does not exceed 0.14 m^3 (5 ft^3).

14.9.3 Solid Fuel Handling and Ash Removal.

14.9.3.1 Solid fuel shall be ignited with a

match or an approved built-in gas flame or other approved ignition source.

14.9.3.2 Combustible or flammable liquids shall not be used to assist ignition.

14.9.3.3 Matches and other portable ignition sources shall not be stored in the vicinity of the solid fuel appliance.

14.9.3.4 Solid fuel shall be added to the fire as required in a safe manner and in quantities and ways not creating a higher flame than is required.

14.9.3.5 Long-handled tongs, hooks, and other required devices shall be provided and used in order to safely add fuel, adjust the fuel position, and control the fire without the user having to reach into the firebox.

14.9.3.6 Ash Protection.

14.9.3.6.1 Ash, cinders, and other fire debris shall be removed from the firebox at adequately regular intervals to prevent interference with the draft to the fire and to minimize the length of time the access door is open.

14.9.3.6.2 All ash shall be removed from the chamber a minimum of once a day.

14.9.3.6.3 The ash shall be sprayed adequately with water before removal in order to extinguish any hot ash or cinders and to control the dust when the ash is moved.

14.9.3.7 Hose Protection.

14.9.3.7.1 For the purposes described in 14.9.3.6.3, and to cool a fire that has become too hot and to stop all fire before the premises is vacated, a water supply with a flexible hose shall be provided at the solid fuel appliance.

14.9.3.7.2 For appliances with fireboxes not exceeding 0.14 m^3 (5 ft^3), the water source shall be permitted to be a 37.9 L (10 gal) container with a gravity arrangement or a hand pump for pressure.

14.9.3.7.3 For appliances with fireboxes over 0.14 m^3 (5 ft^3), the water source shall be a fixed pipe water system with a hose of adequate length to reach the combustion and cooking chambers of the appliance.

14.9.3.7.4 For either applicant, the nozzle shall be fitted with a manual shutoff device and shall be of the type to provide a fine to medium spray of adequate length to reach all areas of the combustion and cooking chambers.

14.9.3.7.4.1 The nozzle shall be of the type that cannot produce a straight stream.

14.9.3.8 Ash Removal Container or Cart.

14.9.3.8.1 A heavy metal container or cart (minimum 16 gauge) with a cover shall be provided for the removal of ash.

14.9.3.8.2 The ash removal container or cart shall not exceed a maximum of 75.7 L (20 gal) capacity, shall be assigned for this one purpose, shall be able to be handled easily by any employee assigned the task, and shall pass easily through any passageway to the outside of the building.

14.9.3.8.3 The container or cart shall always be covered when it is being moved through the premises.

14.9.3.8.4 When any hole occurs in a container from corrosion or damage, the container shall be repaired or replaced immediately.

14.9.3.9 Ash Removal Process.

14.9.3.9.1 Tools shall be provided so that ash removal can be accomplished without having to reach into the chamber.

14.9.3.9.2 The ash shall be spread out gently in small lots on the chamber floor or on a shovel to be sprayed before it is removed to the metal container or cart.

14.9.3.9.3 If the floor of the chamber is of a metal that is subject to rapid corrosion from water, then a non-combustible, corrosion-resistant pan shall be placed just outside the clean-out door for this purpose.

14.9.3.9.4 The ash shall be carried to a separate heavy metal container (or dumpster) used exclusively for the purpose.

14.9.4 Other Safety Requirements.

14.9.4.1 Metal-fabricated solid fuel cooking appliances shall be listed for the application where produced in practical quantities or shall be approved by the authority having jurisdiction.

14.9.4.2 Where listed, they shall be installed in accordance with the terms of their listing and with the applicable requirements of this standard.

14.9.4.3 Site-Built Solid Fuel Cooking Appliances.

14.9.4.3.1 Site-built solid fuel cooking appliances shall be submitted for approval to the authority having jurisdiction before being considered for installation.

14.9.4.3.2 All units submitted to the authority having jurisdiction shall be installed, operated, and maintained in accordance with the approved terms of the manufacturer's instructions and any additional requirements set forth by the authority having jurisdiction.

14.9.4.4 Except for the spark arresters required in 14.1.6, there shall be no additional devices of any type in any portion of the appliance, flue pipe, and chimney of a natural draft solid fuel operation.

14.9.4.5 No solid fuel cooking device of any type shall be permitted for deep-fat frying involving more than 0.95 L (1 qt) of liquid shortening, nor shall any solid fuel cooking device be permitted within 0.92 m (3 ft) of any deep-fat frying unit.

Chapter 15

Down Draft Appliance Ventilation Systems

15.1* General Requirements. Down draft appliance ventilation systems containing or for use with appliances used in processes producing smoke or grease-laden vapors shall be equipped with components complying with the following:

(1) The clearance requirements of Section 4.2
(2) The hood portion complying with the requirements of Chapter 5
(3) Grease removal devices complying with Chapter 6
(4) Special-purpose filters as listed in accordance with UL 1046
(5) Exhaust ducts complying with Chapter 7
(6) The air movement requirements of 8.2.1.2 and 8.2.2.3
(7) Auxiliary equipment (such as particulate and odor removal devices) complying with Chapter 9
(8) Fire-extinguishing equipment complying with the requirements of Chapter 10 and as specified in Section 15.2
(9) The use and maintenance requirements of Chapter 11
(10) The minimum safety requirements of Chapter 12

15.2 Fire-Extinguishing Equipment. For fire-extinguishing equipment on down draft appliance ventilation systems, the following shall apply:

(1) Cooking surface protection shall be provided.
(2) At least one fusible link or heat detector shall be installed within each exhaust duct opening in accordance with the manufacturer's listing.
(3) A fusible link or heat detector shall be provided above each protected cooking appliance and in accordance with the extinguishing system manufacturer's listing.
(4) A manual activation device shall be provided as part of each appliance at a height acceptable to the authority having jurisdiction.

15.2.1 A listed down draft appliance ventilation system employing an integral fire-extinguishing system including detection systems that has been evaluated for grease and smoke capture, fire extinguishing and detection shall be considered as complying with Section 15.2.

15.2.2 The down draft hood system shall be provided with interlocks such that cooking fuel supply will not be activated unless the exhaust and supply air systems have been activated.

15.3 Airflow Switch or Transducer.

15.3.1 An airflow switch or transducer shall be

provided after the last filter component to ensure that a minimum airflow is maintained.

15.3.2 The airflow switch or transducer shall open the interlock circuit when the airflow falls 25 percent below the system's normal operating flow or 10 percent below its listed minimum rating, whichever is lower.

15.3.3 The airflow switch or transducer shall be a manual reset device or circuit.

15.4 Surface Materials. Any surface located directly above the cooking appliance shall be of non-combustible or limited-combustible materials.

Chapter 16
NFPA 17 Standard for Dry Chemical Extinguishing Systems
2002 Edition

NFPA 17 is the standard for dry chemical extinguishing systems that encompasses all types of applications and the requirements of those applications. Chapter 9 of this standard deals with Pre-Engineered Systems. Most restaurant fire suppression systems fall within this type.

Chapter 9 - Pre Engineered Systems

9.1 Uses.

9.1.1* Pre-engineered systems shall be

installed to protect hazards within the limitations of the listing.

9.1.2 Fire-extinguishing systems referenced in 9.1.1 shall comply with UL 1254, Pre-Engineered Dry Chemical Extinguishing System Units, or equivalent listing standard.

9.1.3 Only system components referenced in the manufacturer's listed installation and maintenance manual or alternative suppliers' components that are listed for use with the specific extinguishing system shall be used.

9.2 Types of Systems. Pre-engineered dry chemical systems shall be of the following types:
1. (1) Local application
2. (2) Total flooding
3. (3) Hand hose line
4. (4) Combination of local application and total flooding

9.3 Restaurant Hood, Duct, and Cooking Appliance Systems.

9.3.1 Each protected cooking appliance(s), individual hood(s), and branch exhaust duct(s) directly connected to the hood shall be protected by a single system or by systems designed for simultaneous operation.
1. **9.3.1.1** At least one fusible link or heat detector shall be installed within each exhaust duct opening in accordance with the manufacturer's listing.
2. **9.3.1.2** A fusible link or heat detector shall be provided above each protected cooking appliance and in accordance with the extin-

guishing system manufacturer's listing.

3. **9.3.1.3** Fusible links or heat detectors located at or within 12 in. (305 mm) into the exhaust duct opening and above the protected appliance shall be permitted to meet the requirements of 9.3.1.2.

4. **9.3.2** Fire-extinguishing systems referenced in 9.3.1 shall comply with UL 300, Fire Extinguishing Systems for Protection of Restaurant Cooking Areas, Fire Testing of, or equivalent listing standard.

2002 Edition

1. **9.3.2.1** Fixed automatic dry chemical extinguishing systems shall be installed in accordance with the terms of the listing, the manufacturer's instructions, and this standard.

2. **9.3.3** Systems protecting two or more hoods or plenums, or both, that meet the requirements of Section 5.2 shall be installed to ensure the simultaneous operation of all systems protecting the hoods, plenums, or both, and associated cooking appliances located below the hoods.

9.3.4* Protection of Common Exhaust Duct.

9.3.4.1 Common exhaust ducts shall be protected by one of the following methods:

(1)* Simultaneous operation of all independent hood, duct, and appliance protection systems
(2)* Simultaneous operation of any hood, duct and appliance protection system and the system(s) protecting the entire common exhaust duct.

1. **9.3.4.1.1** A fusible link or heat detector shall be located at each branch duct-to-common duct connection.
2. **9.3.4.1.2** Actuation of any branch duct-to-common duct fusible link or heat detector shall actuate the common duct system only, or when all systems are connected to a control panel in accordance with NFPA 72, National Fire Alarm Code.
3. **9.3.4.2** All sources of fuel or heat to appliances served by the common exhaust duct shall be shut down upon actuation of any protection system in accordance with 9.3.5.
4. **9.3.4.3** The building owner(s) shall be responsible for the protection of a common exhaust duct(s) used by more than one tenant.
1. **9.3.4.3.1** The tenant shall be responsible for the protection of a common exhaust duct(s) serving hoods located within the tenant's space and up to the point of connection to the building owner's common exhaust duct.
2. **9.3.4.3.2** The tenant's common duct shall be considered a branch duct to the building owner's common duct.

9.3.5* Shutoff Devices. Upon activation of any cooking equipment fire-extinguishing system, all sources of fuel and electric power that produce heat to all equipment protected by the system shall be shut down.

9.3.5.1 Steam supplied from an external source does not require shutdown.

9.3.5.2* Exhaust fans and dampers are not required to be shut down upon system actuation.

1. **9.3.5.3** Any gas appliance not requiring protection but located under the same ventilating equipment shall be automatically shut off upon actuation of any extinguishing system.

2. **9.3.5.4** Shutoff devices shall require manual resetting prior to fuel or power being restored.

3. **9.4 Manual Activation Requirements.**

1. **9.4.1** Such mechanical means shall not rely on any of the hardware components that would be common to the automatic function of the fixed fire-extinguishing equipment.

2. **9.4.2** The means for manual actuator(s) actuation shall be mechanical and shall not rely on electrical power for actuation.

2002 Edition

1. **9.4.3** Electrical power shall be permitted to be used for manual actuation if a reserve power supply is provided in accordance with Section 9.6.

2. **9.4.4** The manual actuation means of an automatic extinguishing system shall be totally independent of the automatic means.

3. **9.4.5** A failure of a system component shall not impair both the automatic and manual means of actuation.

4. **9.5 System Annunciation Requirements.**

1. **9.5.1** Upon actuation of a fixed automatic fire-extinguishing system, an audible alarm or visual indicator shall be provided to show that the system has actuated.

2. **9.5.2** Where a fire alarm signaling system is serving the occupancy where the extinguishing system is located, the actuation of the fixed automatic fire-extinguishing system shall actuate the fire alarm signaling system.

3. **9.6 System Supervision.**

1. **9.6.1** Where electrical power is required to operate the fixed automatic fire-extinguishing system, it shall be monitored by a supervisory alarm, with a reserve power supply provided.

2. **9.6.2** Where fixed automatic fire-extinguishing systems include automatic mechanical detection and actuation as a backup detection system, electrical power monitoring is not required.

3. **9.6.3** Where fixed automatic fire-extinguishing systems are interconnected or interlocked with the cooking equipment power sources so that if the fire-extinguishing system becomes inoperable due to power failure, all sources of fuel and heat to all cooking equipment serviced by that hood shall automatically shut off and electrical power monitoring is not required.

4. **9.7 Review and Certification.**

9.7.1 If required, compete drawings of the system installation shall be submitted to the authority having jurisdiction.

9.7.1.1 System drawings shall include the following:
1. (1) Hood(s)
2. (2) Exhaust duct(s)
3. (3) Appliances
4. (4) Interface of the fire-extinguishing system detectors
5. (5) Piping
6. (6) Nozzles
7. (7) Fuel shutoff devices
8. (8) Agent storage container(s)
9. (9) Manual actuation device(s)

9.7.2* Design and installation of systems shall be performed only by persons properly trained and qualified to design and/or install the specific system being provided.

9.7.2.1 The installer shall provide certification to the authority having jurisdiction that the installation is in complete agreement with the terms of the listing and the manufacturer's instructions and/or approved design.

Chapter 17

Standard for Wet Chemical Extinguishing Systems 2002 Edition

Chapter 4 Components

4.1 General. Only system components referenced or permitted in the manufacturer's listed installation and maintenance manual or alternate components that are listed for use with the specific extinguishing system shall be used.

4.2 Detectors. Detectors shall be listed or approved devices that are capable of detecting heat.

4.3 Discharge Nozzles. See Section 5.5.

4.3.1 Discharge nozzles shall be listed for their intended use.

4.3.1.1 Discharge nozzles shall be provided with an internal strainer or a separate listed strainer located immediately upstream of the nozzle.

4.3.1.2 Discharge nozzles shall be of brass, stainless steel, or other corrosion-resistant materials, or be protected inside and out against corrosion.

4.3.1.3 Discharge nozzles shall be made of non-combustible materials and shall withstand the expected fire exposure without deformation.

2002 Edition

4.3.1.4* Discharge nozzles shall be permanently marked for identification.

4.3.1.5 All discharge nozzles shall be provided with caps or other suitable devices to prevent the entrance of grease vapors, moisture, or other foreign materials into the piping.

4.3.1.6 The protection device shall blow off, open, or blow out upon agent discharge.

4.4 Operating Devices

4.4.1 Operating devices shall be listed.

4.4.1.1 Operating devices shall be designed for the service they will encounter and shall not be rendered inoperative or susceptible to accidental operation.

4.4.1.2 Operating devices shall be designed to function properly through a minimum temperature range from 32°F to 120°F(0°C to 49°C)or marked to indicate their temperature limitations.

4.4.2 Manual Actuators.

4.4.2.1 Manual actuators shall not require a force of more than 40 lb (178N).

4.4.2.2 Manual actuators shall not require a movement of more than 14 in. (356 mm) to secure operation.

4.4.2.3 All manual actuators shall be provided with operating instructions.

4.4.2.4 These instructions shall be permitted to include the use of pictographs and shall have lettering at least 4 in. (6.35 mm) in height. (See 5.2.1.4.)

4.4.2.5 All remote manual operating devices shall be identified as to the hazard they protect.

4.4.3 Shutoff Devices

4.4.3.1 On activation of any cooking equipment fire-extinguishing system, all sources of fuel and electric power that produce heat to all equipment protected by the system shall be shut down.

4.4.3.2 Gas appliances not requiring protection but located under the same ventilation equipment shall also be shut off.

4.4.3.3 Steam supplied from an external source shall not be required to be shut down.

4.4.3.4 Solid fuel cooking operations shall not be required to shut down.

4.4.3.5 Exhaust fans and dampers are not required to be shut down on system actuation as the systems have been tested under both zero-and high-velocity flow conditions.

4.4.3.6 If the expellant gas is used to pneumatically operate these devices, the gas shall be taken prior to its entry into the wet chemical tank.

4.4.3.7 Shutoff devices shall require manual resetting prior to fuel or power being restored.

4.5 Pipe and Fittings, Tubing, Hose.

4.5.1* General. Pipe and associated fittings shall be of non-combustible material having physical and chemical characteristics compatible with the wet chemical solution.

4.5.2 Galvanized pipe and fittings shall not be used unless specifically listed with the system.

4.5.3 The pressure rating of the pipe fittings and connection joints shall withstand the maximum expected pressure in the piping system.

4.5.4 Pipe, tubing, hose, and fitting materials and types shall be in accordance with the manufacturer's listed installation and maintenance manual.

4.6 Wet Chemical.

4.6.1* The type of wet chemical used in the system shall be listed for the particular system and recommended by the manufacturer of the wet chemical system.

4.6.2 Wet chemical solutions of different formulations or different manufacturers shall not be mixed.

4.7 Electrical Wiring and Equipment. Electrical wiring and equipment shall be installed in accordance with NFPA 70, National Electrical Code, or the requirements of the authority having jurisdiction.

4.8 Indicators. Wet chemical systems shall be provided with an audible or visual indicator to show that the system is in a ready condition or is in need of recharging.

Chapter 5 System Requirements

5.1 General.

5.1.1 Wet chemical fire-extinguishing systems shall comply with standard UL 300, Fire Testing of Fire Extinguishing Systems for Protection of Restaurant Cooking Areas.

5.1.2 Use. Hazards and equipment that can be protected using wet chemical extinguishing systems include the following:

(1) Restaurant, commercial, and institutional hoods
(2) Plenums, ducts, and filters with their associated cooking appliances
(3) Special grease removal devices
(4) Odor control devices
(5) Energy recovery devices installed in the exhaust system

5.1.3 Applications. The manufacturer's listed installation and maintenance manual shall be consulted for system limitations and applications for which wet chemical extinguishing systems are considered satisfactory protection.

5.1.4 Each protected cooking appliance, individual hood, and branch exhaust duct directly connected to the hood shall be protected by a system or systems designed for simultaneous operation.

5.1.5 Where two or more hazards can be simultaneously involved in fire by reason of their proximity, the hazards shall be protected by either of the following:
(1) Individual systems installed to operate simultaneously.
(2) A single system designed to protect all hazards that can be simultaneously involved.

5.1.5.1 Any hazard that will allow fire propagation from one area to another shall constitute a single fire hazard.

5.2 System Actuation.

5.2.1 All systems shall have both automatic and manual methods of actuation.

5.2.1.1 The automatic and manual means of system actuation, external to the control head or releasing device, shall be **SYSTEM REQUIREMENTS 17A-7**, separate and independent of each other so that a failure of one will not impair the operation of the other.

5.2.1.2 When a listed releasing mechanism is used employing a single line for mechanical detection and remote manual control, the remote manual control shall be installed inline, prior to all detection devices, so malfunction of one does not impede operation of the other.

5.2.1.3 Automatic detection and system actuation shall be in accordance with the manufacturer's listed installation and maintenance manual.

5.2.1.4 All devices necessary for proper operation of the system shall function simultaneously with the system operation.

5.2.1.5 Operation of any manual actuator shall be all that is required to bring about the full operation of the system.

5.2.1.6 At least one manual actuator shall be provided for each system.

5.2.1.7 All operating devices shall be designed, located, installed, or protected so that they are not subject to mechanical, environmental, or other conditions that could render them inoperative or cause inadvertent operation of the system.

5.2.1.8 An audible or visual indicator shall be provided to show that the system has operated, that personnel response is needed, and that the system is in need of recharge.

5.2.1.9 The extinguishing system shall be connected to the fire alarm system, if provided, in accordance with the requirements of NFPA 72, National Fire Alarm Code, so that the actuation of the extinguishing system will sound the fire alarm as well as provide the function of the extinguishing system.

5.2.1.10* A readily accessible means for manual activation shall be located in a path of egress. When manual activation is used for cooking related protection, the manual activation device shall be installed no more than 48 in. (1200 mm), nor less than 42 in. (1067 mm) above the floor and shall clearly identify the hazard protected.

5.2.1.11 Automatic systems protecting common exhaust ducts only shall not require a remote manual actuator.

5.2.1.12 The means for mechanical actuator(s) shall be mechanical and shall not rely on electrical power for actuation.

5.2.1.13 Electrical power shall be permitted to be used for manual activation if a reserve power supply is provided or if supervision is provided as per Section 5.3.

5.3 Supervision

5.3.1 Where supervision of any or all of the following is provided, it shall be designed to give an indication of trouble in the following:

(1) Automatic detection system
(2) Electrical actuation circuit
(3) Electrical power supply

5.3.2 Signals indicating the failure of supervised devices or equipment shall give prompt and positive indication of any failure and shall be distinctive from signals indicating operation or hazardous conditions.

5.4 System Location.

5.4.1 Wet chemical containers and expellant gas assemblies shall be located within the temperature range specified in the manufacturer's listed installation and maintenance manual.

5.4.2 If ambient temperatures outside the manufacturer's operating temperature range are expected, protection shall be provided to maintain the temperature within the listed range.

5.4.3 Wet chemical containers and expellant gas assemblies shall not be located where they could be subjected to mechanical, chemical, or other damage.

5.4.4 Where damage due to chemical or mechanical exposure is expected, protective devices such as enclosures or guards acceptable to the authority having jurisdiction shall be provided.

5.4.5 Wet chemical containers and expellant gas assemblies shall be accessible for inspection, maintenance, and recharge.

5.4.6 Wet chemical containers and expellant gas assemblies shall be located near the hazard or hazards protected but not where they will be exposed to the fire.

5.5 Discharge Nozzles. See Section 4.3.

5.5.1 All discharge nozzles shall be designed and subsequently located, installed, or protected so that they are not subject to mechanical, environmental, or other conditions that could render them inoperative.

5.5.2 Discharge nozzles shall be connected and supported.

5.6 Special Requirements.

5.6.1 Systems protecting two or more hoods or plenums, or both, that meet the requirements of 5.1.5 shall be installed to ensure the simultaneous operation of all systems protecting the hoods, plenums, and associated cooking appliances located below the hoods.

5.6.1.1 The building owner(s) shall be responsible for the protection of a common exhaust duct(s) used by more than one tenant.

5.6.1.2 The tenant shall be responsible for the protection of common exhaust duct(s) serving hoods located within the tenant's space and up to the point of connection to the building owner's common exhaust duct.

5.6.1.3 The tenant's common duct shall be considered a branch duct to the building owner's common duct.

5.6.1.4 A single listed detection device shall be permitted for more than one appliance when installed in accordance with the system's listing.

5.6.1.5 At least one fusible link or heat detector shall be installed within each exhaust duct opening in accordance with the manufacturer's listing.

5.6.1.6 A fusible link or heat detector shall be provided above each protected cooking appliance and in accordance with the extinguishing system manufacturer's listing.

5.6.1.6.1 Fusible links or heat detectors located at or within 12 in. (305 mm) into the exhaust duct opening and above the protected appliance shall be permitted to meet the requirements of 5.6.1.6.

5.6.1.7 Where the pipe or other conduit penetrates a duct or hood, the penetration shall have a liquid tight continuous external weld or shall be sealed by a listed device.

5.6.2* Protection of Common Exhaust Duct.

5.6.2.1 Common exhaust ducts shall be protected by one of the following methods:

(1) Simultaneous operation of all independent hood, duct, and appliance protection systems.
(2) Simultaneous operation of any hood, duct, and appliance protection system and the system(s) protecting the entire common exhaust duct

5.6.2.1.1 A fusible link or heat detector shall be located at each branch duct-to-common duct connection.

5.6.2.1.2 Actuation of any branch duct-to-common exhaust duct fusible link or heat detector shall actuate the common duct system only, or when all independent systems are connected to a control panel in accordance with NFPA 72, National Fire Alarm Code.

5.6.2.2 All sources of fuel or heat to appliances served by the common exhaust duct shall be shut down upon actuation of any protection system in accordance with 4.4.3.

5.6.3* Ignition sources contained within any exhaust system shall be protected and have a separate detection system in accordance with the manufacturer's recommendations and that is approved by the authority having jurisdiction.

5.6.4 Movable cooking equipment shall be provided with a means to ensure that it is correctly positioned in relation to the appliance discharge nozzle during cooking operations.

Chapter 6 Plans and Acceptance Tests

6.1* Specifications. Specifications for wet chemical fire-extinguishing systems shall be drawn up with care under the supervision of a trained person and with the advice of the authority having jurisdiction.

6.1.1 The following items shall be included in the specifications:

(1) Designation of the authority having jurisdiction and indication of whether plans are required.

(2) Statement that the installation conforms to this standard and meets the approval of the authority having jurisdiction.

(3) Indication that only equipment that is referenced in the manufacturer's listed installation and maintenance manual or alternate suppliers' components that are listed for use with the specific extinguishing system shall be used.

(4) Identification of special auxiliary devices acceptable to the system manufacturer and the authority having jurisdiction.

(5) List of the specific tests, if any, that are required.

(6) Identification of the hazard to be protected and including such information as physical dimensions, cooking appliances, energy sources for each appliance, and air-handling equipment.

6.2 Review and Certification. Design and installation of systems shall be performed only by persons properly trained and qualified to design and/or install the specific system being provided. The installer shall provide certification to the authority having jurisdiction that the installation is in complete agreement with the terms of the listing and the manufacturer's instructions and/or approved design.

6.3 Plans. Where plans are required, the responsibility for their preparation shall be entrusted only to trained persons.

2002 Edition

6.3.1 The plans shall be drawn to an indicated scale or be suitably dimensioned and shall be reproducible.

6.3.2 The plans shall contain sufficient detail to enable the authority having jurisdiction to evaluate the protection of the hazard(s).

6.3.3 The details on the system shall include the following:

(1) Size, length, and arrangement of connected piping
(2) Description and location of nozzles

6.3.4 Information shall be submitted pertaining to the following:

(1) The location and function of detection devices
(2) Operating devices
(3) Auxiliary equipment
(4) Electrical circuitry

6.3.5 Approval of Plans. Where plans are required, they shall be submitted to the authority having jurisdiction for approval before work starts.

6.3.6 Where field conditions necessitate any substantial change from the approved plan, the as-installed plans shall be submitted to the authority having jurisdiction for approval.

6.4 Approval of Installations. The completed system shall be tested by trained personnel as required by the manufacturer's listed installation and maintenance manual.

6.4.1 The tests shall determine that the system has been properly installed and will function as intended.

6.4.2 The installer shall certify that the system has been installed in accordance with the approved plans and the manufacturer's listed installation and maintenance manual.

6.4.3 Where required by the authority having jurisdiction, the approval tests shall include a discharge of wet chemical to verify that the system is properly installed and functional.

6.4.4 The owner shall be provided with a copy of the manufacturer's listed installation and maintenance manual or listed owner's manual.

Chapter 7 Inspection, Maintenance, and Recharging

7.1 General.

7.1.1* Storage. Recharging supplies of wet chemical shall be stored in the original closed shipping container supplied by the manufacturer.

7.1.1.1 These containers shall not be opened until the system is recharged.

7.1.1.2 Wet chemical supplies shall be maintained within the manufacturer's recommended storage temperature range.

7.1.2 Expellant Gas. A method and instructions shall be provided for checking the amount or the pressure of expellant gas to ensure that it is sufficient for the proper operation of the system.

7.1.3 Access. System access for inspection or maintenance that requires opening panels in fire chases or ducts, or both, shall not be permitted while any appliance(s) or equipment protected by that system is in operation.

7.1.4* Recharge. After any discharge, or if insufficient charge is noted during an inspection or maintenance procedure, the following procedures shall be conducted in accordance with the manufacturer's listed installation and maintenance manual:

(1) The system shall be properly recharged.
(2) The system shall be placed in the normal operating condition.
(3) The piping shall be flushed in accordance with the manufacturer's recommended instructions (only following a discharge).

7.2 Owner's Inspection.

7.2.1 Inspection shall be conducted on a monthly basis in accordance with the manufacturer's listed installation and maintenance manual or the owner's manual.

7.2.2 Inspections shall include verification of the following:

(1) The extinguishing system is in its proper location.
(2) The manual actuators are unobstructed.
(3) The tamper indicators and seals are intact.
(4) The maintenance tag or certificate is in place.
(5) No obvious physical damage or condition exists that might prevent operation.
(6) The pressure gauge(s), if provided, is in operable range.
(7) The nozzle blowoff caps are intact and undamaged.
(8) The hood, duct, and protected cooking appliances have not been replaced, modified or relocated.

7.2.3 If any deficiencies are found, appropriate corrective action shall be taken immediately.

7.2.4 Personnel making inspections shall keep records for those extinguishing systems that were found to require corrective actions.

7.2.5 At least monthly, the date the inspection is performed and the initials of the person performing the inspection shall be recorded.

7.2.6 The records shall be retained for the period between the semiannual maintenance inspections.

7.3 Maintenance.

7.3.1 A trained person who has undergone the instructions necessary to perform the maintenance and recharge service reliably and has the applicable manufacturer's listed installation and maintenance manual and service bulletins shall service the wet chemical fire - extinguishing system 6 months apart as outlined in 7.3.2.

7.3.2 At least semiannually, maintenance shall be conducted in accordance with the manufacturer's listed installation and maintenance manual.

7.3.2.1 Maintenance shall include the following:

(1) A check to see that the hazard has not changed.
(2) An examination of all detectors, the expellant gas container(s), the agent containers(s), releasing devices, piping, hose assemblies, nozzles, signals, all auxiliary equipment, and the liquid level of all non-pressurized wet chemical containers.
(3) Verification that the agent distribution piping is not obstructed.

7.3.2.2 Where semiannual maintenance of any wet chemical containers or system components reveals conditions such as, but not limited to, corrosion or pitting in excess of the manufacturer's limits; structural damage or fire damage; or repairs by soldering, welding, or brazing; the affected part(s) shall be replaced or hydrostatically tested in accordance with the recommendations of the manufacturer or the listing agency.

7.3.2.3 All wet chemical systems shall be tested, which shall include the operation of the detection system signals and releasing devices, including manual stations and other associated equipment.

7.3.2.4 Where the maintenance of the system(s) reveals defective parts that could cause an impairment or failure of proper operation of the system(s), the affected parts shall be replaced or repaired in accordance with the manufacturer's recommendations.

7.3.2.5 The maintenance report, with recommendations, if any, shall be filed with the owner or with the designated party responsible for the system.

7.3.2.6 Each wet chemical system shall have a tag or label securely attached, indicating the month and year the maintenance is performed and identifying the person performing the service. Only the current tag or label shall remain in place.

7.3.3 Fixed temperature-sensing elements of the fusible metal alloy type shall be replaced at least annually from the date of installation. They shall be destroyed when removed.

7.3.3.1 The year of manufacture and the date of installation of the fixed temperature-sensing element shall be marked on the system inspection tag. The tag shall be signed or initialed by the installer.

7.3.4 Fixed temperature-sensing elements other than the fusible metal alloy type shall be permitted to remain continuously in service, provided they are inspected and cleaned or replaced if necessary in accordance with the manufacturer's instructions every 12 months or more frequently to ensure proper operation of the system.

CHAPTER 18

UL 300, Manufacturer's Pre-Engineered Manual and UL Listing

Let's talk about UL 300 and what it is. UL 300, *Standard For Fire Testing of Fire Extinguishing Systems for Protection of Restaurant Cooking Areas,* is the protocol set forth by Underwriters Laboratories in 1992, with a Second Edition in July 1996, that caused the testing requirements for cooking area suppression systems to be changed.

What prompted the change in testing protocol? Well, there were numerous reasons such as America becoming more health conscious, resulting in commercial cooking operations to change from animal fat oils to more non-fat oils such as vegetable or soy. The lack of fat in the oils became a problem for the dry chemical system to saponify the oil. Saponification is a chemical reaction that occurs between the cooking oil and dry chemical powder. Soapy foam is created at the surface of the oil not allowing for the introduction of oxygen and thus the tetrahedron collapses or in laymen terms-the fire is extinguished.

In addition, to keep auto ignition from reoccurring a cooling process must begin to bring the oil below its lower flammable limit. Manufacturers have begun to insulate appliances to allow better or more efficient use of the cooking energy source. This containment of BTUs being generated from the appliance had a reverse effect on the saponification. This lowering of the temperature rate was not occurring rapidly enough before auto ignition reoccurred.

The wet chemical solutions that utilize a high alkalinity base were able to pass the new requirements and the dry chemicals could not. Understand, the dry chemicals did not lose their UL listings provided the manufacturers still supported the product. Basically, the dry chemical systems were grand-fathered in.

The UL Listing of the system incorporates NFPA 96, 17 or 17A and the manufacturer's manual. It is critical that the manufacturer's installation and maintenance be adhered to during the installation of the system and the maintenance of the suppression system.

CHAPTER 19

Class K

For the reasons stated in Chapter 18, UL and NFPA 10 have designed a new class for "cooking grease" fires, Class K. As a back up for the dry chemical suppression systems, a Class K extinguisher is required in the kitchen area. A Class K extinguisher basically looks like a water extinguisher that is filled with the same type of wet chemical agent found in the wet suppression systems.

CHAPTER 20

Investigation of Commercial Cooking Fires

BEGINNING THE INVESTIGATION:

You should now have a basic understanding of the requirements for the installation of the ventilation system and fire suppression system.

Remember, if a fire occurs within the protected area of the fire suppression system and the extinguishing of the fire does not occur- **SOMETHING IS WRONG!** It then becomes a matter of finding out what.

If a suppression system fails to extinguish a fire within the confines of the ventilation system, the fire will probably extend into the roof line and a total loss of the building could occur. For this reason restaurant fires are usually a major loss. You as the Investigator for the restaurant as well as the additional Origin and Cause Investigators, may be in the position of exposure if the building owner is not able to retrieve the evidence necessary for further examination. It takes a lot of trained manpower and the proper equipment to complete this task. Always make sure all evidence that is requested is retained. I personally take what I call "soup to nuts" - the entire cooking line, ventilation system and suppression system. This of course may vary with each investigation and the severity of the loss.

After the O&C Investigator has conducted a separate origin and causation investigation, then the examination of the ventilation and suppression system can commence. The investigation should start as any origin and cause investigation does. Basic Methodology as stated within NFPA 921, *Guide for Fire Explosion Investigations*, must be adhered to. Once an area of origin or point of origin is determined, such as a fire within a fryer vat, the progression of said fire should be documented. Most cooking fires are witness events but don't be surprised that the extension of the fire within the ventilation system was not observed. Most interviews or fact depositions that I have reviewed indicate that the restaurant employee is so intent on the extinguishing of the flame coming from a vat of oil or other cooking device that the extension of the flame against the grease filters is not noticed. It is only when a police officer or civilian sees the fire coming out the exhaust fan and advises the restaurant of the fire or when the unusual vibrations or noises from the ventilation system are heard do the restaurant employees become aware of the fire.

After your interviews are conducted and the scene is photographed as found, a reconstruction (if possible) of the ventilation system, suppression system and cooking line should occur. I have experienced a restaurant totally destroyed with the cooking line collapsing into the basement. This type of fire damage makes it impossible to reconstruct the cooking line and its ventilation and suppression system within the confines of the building due to safety concerns alone. The next best plan of action is to retrieve the appliances, ventilation system and suppression system, say on the parking lot of the restaurant. You will find a crane or high jacks to be useful for this purpose.

Examination of the Ventilation System

A detailed examination of the ventilation system should be done. Check for the type of grease filters being utilized. Usually the firefighters during suppression activities remove the filters and the exact location of the filters will be unknown. The filters over or next to the area of origin may show signs of oxidation or even melting depending on the type of material that the filter is constructed of. MESH FILTERS are not and have never been approved for cooking use. These filters absorb grease and cause a deep-seated fire to occur. Suppression systems are not tested or designed for these types of fires.

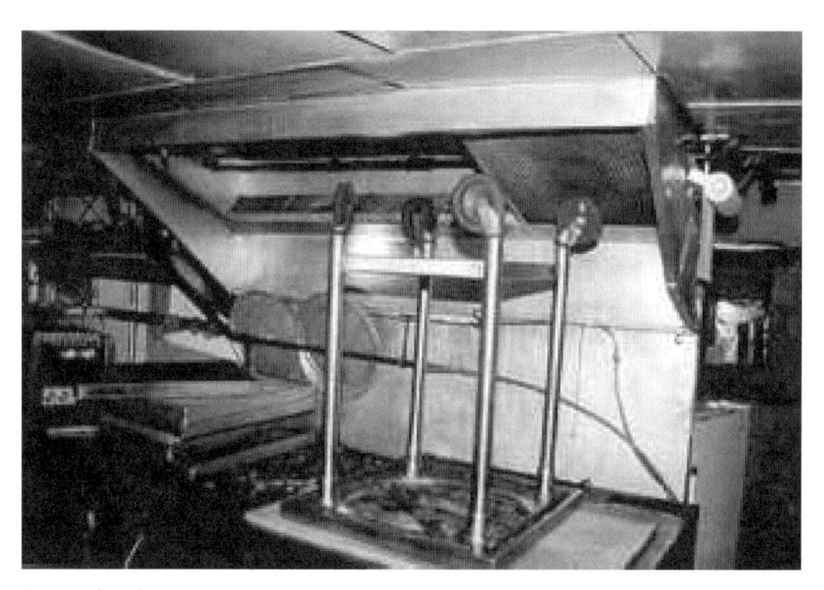

Description;
Mesh Filter

Clearance to combustibles, liquid tight seams, proper gauge of metal and other requirements as referenced in Chapters 3 through 7 should be noted and documented. From experience I have found that the extension of the fire outside of the ventilation system, a large percent of the time, can be attributable to improper clearance to combustibles and non liquid tight seams.

Description;
Improper Clearance to Combustibles

Description; Non-Liquid Tight Penetration

The exhaust fan should be documented and the physical turning of the pulley for the motor and fan should be done to see if any seizing of the fan or bearings occurred. Some Origin and Cause Investigators may place the origin of the fire within the ductwork or an ignition from the exhaust fan. Remember, the exhaust fans are explosion proof and the exhaust ducts are not combustible. The only fuel source other than grease would be creosote in solid fuel cooking. Except for an ember or spark from solid fuel cooking, a source of ignition of cooking grease within the exhaust is needed and usually it is a direct flame impingement from a cooking appliance.

Waterwash hoods are made to extract grease or clean ventilation hoods and do a good job of both if properly maintained. A consult with Underwriters Laboratory indicates that there is not one waterwash hood that is listed to extinguish fires nor is there a suppression system that is listed without duct and plenum protection.

Examination of the Fire Suppression System

The suppression system must be measured in its entirety, not only the discharge piping but also the fuse link line and manual pull station line. Nozzle identification, the distance from the nozzle to appliance coverage and square inches of cooking surfaces being protected must be documented. Also, document if any of the nozzles are clogged.

Description;
Interior View of Nozzles. From Above Fryer

A good way to start the examination of the suppression system would be to do the six-month maintenance procedure as required by the manufacturer. A blow out of the piping after the removal of the nozzles would also be advised. This blow out of the piping can be done by placing balloons on the open end of the pipe where the nozzles would be.

After the balloons have been placed on the piping, air or nitrogen can slowly be applied. A trick of the trade for this procedure is to take a 2 fi gallon water extinguisher, cut the end off of the discharge hose and charge the extinguisher with only nitrogen to its operating pressure. Place the discharge hose of the extinguisher at the end of the main discharge pipe run closest to the suppression system agent cylinder and slowly depress the discharge lever. This will allow you to control the amount of pressure being introduced into the piping. If no obstructions exist, the balloons will slowly expand.

Description;
Balloons on Piping

Description;
Expanding Balloon

I find that placing the suppression system on a CAD design is useful not only to apply the diagram against the manufacturer's requirements for installation but also for explanatory requirements that may be necessary down the road.

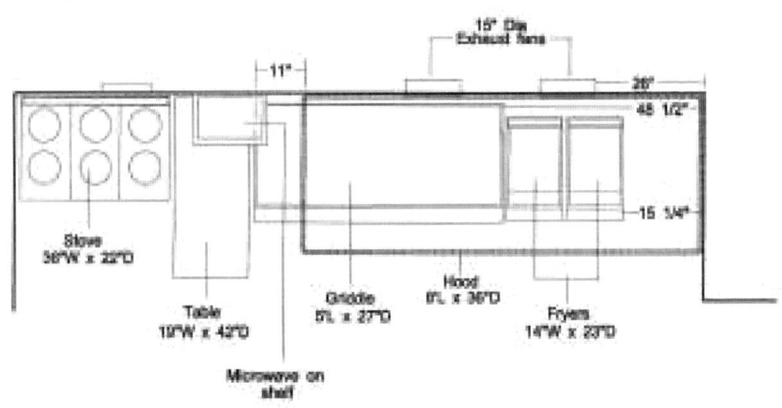

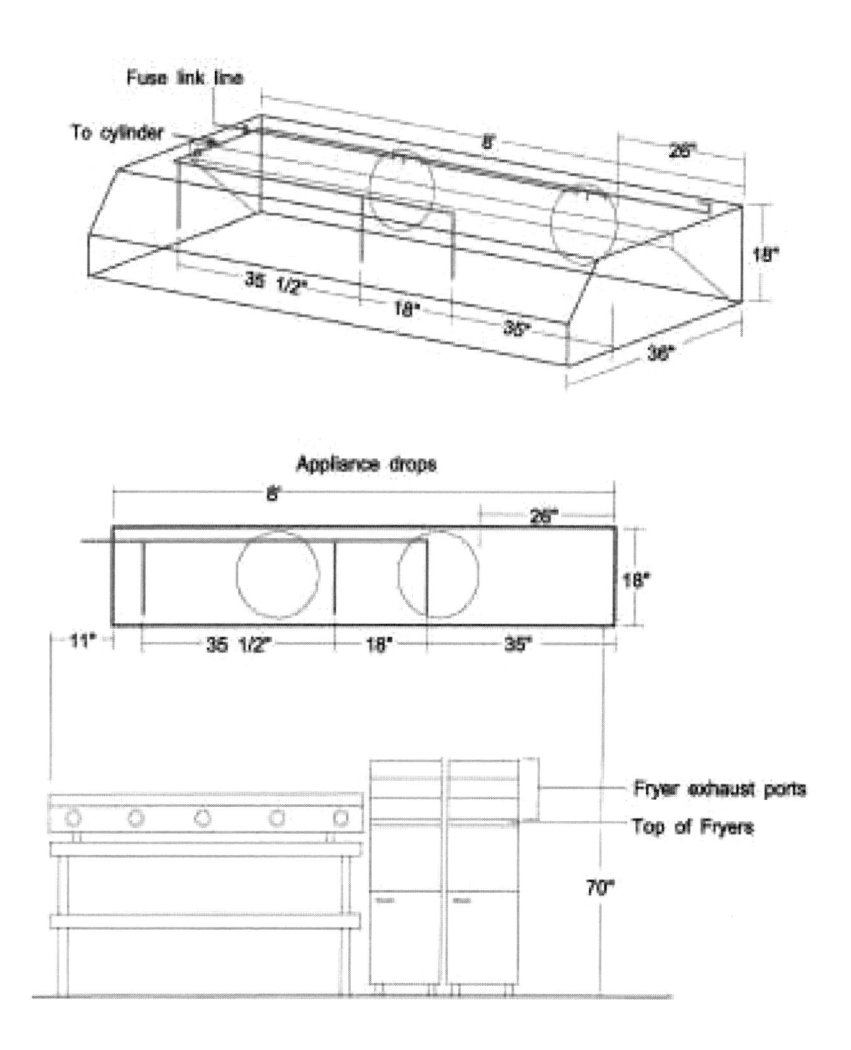

I have also begun to use what I call projection sticks on the nozzles. You may be surprised to find out where the nozzle is actually pointing over the appliance. It must be noted that the nozzle drop must be rigid for this type of examination.

Description;
View(s)of Nozzle Projectory Above Griddle and Fryer

Note which fuse links activated and if the manual pull station was utilized. I have heard arguments over the years that employees did not use the manual pull station during the fire. My knowledge of the Fire Service industry dictates that no matter how many times you teach a civilian what to do in case of fire, you really do not know how they will react. Manual pulls are to be backups to the automatic activation of the suppression system only.

NOTE: Thermal Link Housing not centered in opening of duct

NOTE: Grease caps still on nozzles.

After fully documenting the suppression system, the design and maintenance requirements of the suppression system can now be applied against the manufacturer's manual, as well as standards and codes. The reasons for the lack of extinguishing of the fire by the suppression system should now become apparent.

CHAPTER 21

Retrieval of Evidence

The scene has been documented by all interested parties and a protocol agreed to. Some of the testing may need to be done at a later date, for example, the testing of the regulator on an Ansul system.

All parties must agree on what evidence is being identified and preserved. Each piece of evidence is removed, identified and photographed. A **Transfer of Evidence Form** should be completed.

Transfer of Evidence Form

James F. Valentine Jr. Inc.

FIRE PROTECTION & INVESTIGATIONS

Corporate Office:
P.O. Box 4106
11 N. Berlin Road
Lindenwold, NJ 08021

Tel: (888) 273-6644 Ext. 101
(856) 784-6077 Ext. 101
Fax: (856) 782-7031
Email: [illegible]

EVIDENCE CHANGE OF CUSTODY RECORD

Date: / / Page of

Our File:

The evidence herein described has been transferred on this date to:

Representing: James F. Valentine Jr. Inc. Address: 11 N. Berlin Road, Lindenwold, NJ 08021

From:____________________ Address:____________________

Representing :____________________

Description of evidence:

ITEM# Description Quantity

Signatures:

To:____________________ From:____________________

Date: / / Date: / /

Regional Offices:

Yardley, PA – Philadelphia, PA – Langhorne, PA – North Wales, PA – Jessup, PA – Albany, NY – Riverside, NJ – Garfield, NJ – Wildwood, NJ – Lincoln, DE – Wilmington, DE – Greenville, SC – Raleigh, NC – Knoxville, TN – Glen Allen, VA – Spring Hill, FL

I have found a system where each piece of equipment is photographed with a numerical placard and that number is used as the evidence number for the **Transfer of Evidence Form.**

Again, special equipment such as a plasma cutter may be needed for the proper removal. Damage usually occurs if a demolition company is utilized instead of a trained and qualified retrieval technician. Retrieving evidence is not the same as demolition.

CHAPTER 22

Responsibilities

Responsibilities of the Restaurant Owner

What is the responsibility of the restaurant owner? The answer must be addressed depending on what system we are talking about. If the cleaning of the ventilation system is the question, it would be the responsibility of the restaurant owner to hire a professional to inspect and advise the owner of any required cleaning cycle.

A qualified company that has the necessary expertise to ensure that the system has been installed and maintained in accordance with the manufacturer's requirements must also maintain the suppression system. Again, any recommendations must be noted on the inspection form so that corrected action can be addressed.

Responsibilities of the Ventilation Cleaning Company

The Ventilation Cleaning Company is a professional company hired to inspect and advise the owner of the restaurant of the required cleaning cycle, be it every six months or monthly.

The hood cleaning company must also be relied upon for any other notification of violations that are found during their inspection, such as inaccessible areas of the ventilation system, the need for access panels and all other violations. If the ventilation system is cleaned in accordance with NFPA 96, then no inaccessible areas can exist.

The 2008 edition of NFPA 96, Annex A.11.6.2:

When to clean: A measurement system of deposition should be established to trigger a need to clean, in addition to a time reference based on equipment emissions.

The method of measurement is a depth gauge comb, shown in Figure A.11.6.2, which is scraped along the duct surface. For example, a measurement depth of 2000 mm (0.078 in.) indicates the need to remove the deposition risk.

COMMENT: The addition of this Annex was made so that a physical way to measure grease could be added. The problem becomes, where does an enforcement official (AHJ) obtain such a gauge. The answer my friend is the International Kitchen Exhaust Cleaning Association, located at 100 North 20th Street, Suite 400, Philadelphia, PA 19103. Phone (215) 564-3484.

Responsibilities of the Suppression Company

The Suppression Company is a qualified company that has the necessary licensing and expertise to ensure that the system has been installed and maintained in accordance with the manufacturer's requirements. Any recommendations must be noted on the inspection form so that the necessary actions can be effected. If any violation of the manufacturer's listing is found and is an eminent threat to the extinguishing of a fire within the protected areas of the suppression system, then no inspection tag should be hung on the suppression system until corrective action is taken.

A suppression system maintenance company that has taken over the responsibilities of doing the bi-yearly maintenance procedures but who did not install the suppression system should check the system against the manufacturer's installation requirements. This is to ensure that the system has been installed properly. If you are going to take over the liability for the suppression system then take the time to make sure that the suppression system is in accordance with the manufacturer's listed requirements.

You should then advise the owner of any deficiencies in the suppression system in written form. A good idea would be to note the deficiencies on the inspection report and then send a letter to the client stating the deficiencies in plain English. Also advise the owners of any monthly duties they may be required to perform on the suppression system.

CHAPTER 23

Case Studies

What follows are excerpts of seven case summaries that I have prepared as expert reports for law firms, insurance companies or other principals. Parts of these studies have been edited and/or omitted in the publishing of this book however all relevant text is provided. These reports assert opinions that I have developed from my decades of knowledge, training and experience in fire protection systems, fire science, building construction design, fire reconstruction and fire ground tactical command. I also rely on my personal experience as Chief of the Lindenwold Fire Department in Camden County, New Jersey.
Additional information on my experience, education and qualifications would be provided to the requesting party in my curriculum vitae. Information on other claims in which I provided expert testimony would also be provided, as well as my compensation schedule for background review, report preparation and trial or deposition testimony. In addition, the documents I have reviewed specifically in connection with each case summary are listed as a separate exhibit.

The following case studies are taken from actual investigations that I have conducted. Naturally, the court, jurisdiction, plaintiffs and defendants have been made generic. Names are referenced as "John" or "Jane" Doe. The Suppression System Manufacturer has been identified as "Manufacturer".

CASE STUDY I

DISCUSSION

The origin of this fire was on the natural gas fired char grill located in the southeast corner of the basement kitchen. The cause of this fire has been determined to be accidental ignition of grease and/or vinaigrette sauce during the cooking of porta bella mushrooms on the grill. This normal flare-up extended from the char grill grates, upward to the mesh filters and into the plenum area and ignited grease in the ductwork to the exhaust fan.

This report will address the ventilation and suppression system over the area of origin, specifically the small gas-fired grill located towards the south and east corner of the kitchen workspace.

The grill of origin measured 3′ by 2′ (Exhibit 5) and was located under a 3′ by 30″ ventilation hood. (Exhibit 6) The ventilation hood exhausted from the top rear of the plenum area(area of the hood behind the mesh filter bank and before the duct opening). The 20″ by 8″ exhaust duct traveled horizontally in a southern direction to the south wall of the kitchen where it changed direction and proceeded west for 17.5″ and then went vertical into two 8″ by 10″ ducts. (Exhibit 7) The vertical ducts traveled to the first floor of the building where they connected together in a liquor storage room. The single 20″ by 8″ duct then proceeded through the second floor and into the attic. (Exhibit 8) In the attic the duct again went horizontal and proceeded west for approximately 12 feet and then again went vertical into a 20″ by 20″ duct that exited the building at the roof line to the exhaust fan curbing. (Exhibit 9)

The grill of origin, plenum area of the hood and the duct run was being protected by a wet chemical four gallon extinguishing system. The system protected the grill and ventilation system by use of two appliance nozzles, two duct nozzles and one plenum nozzle. Automatic activation of the suppression system was done by use of a single fuse link located in the area of the top of the hood. Manual activation was accomplished by way of a mechanical pull station located in the flow of egress towards the rear exit. A mechanical gas valve was located in the ceiling of the kitchen. (Exhibit 10) The system was installed by a suppression system company and replaced an existing dry chemical suppression system as evident by the pipe holes left in the hood and duct run. The system was last inspected by_______ in _______.(Exhibit 11) The ventilation system had been manually cleaned in ______ by a ventilation cleaning company. (Exhibit 12) In_________the ventilation system was cleaned by _______ who has cleaned ducts for over twenty years. This change of vendor was due to the fact that ____ decided not to service any more accounts located in this area.

An employee of the restaurant, a Mr. Doe, states in his deposition that he understands the concept of the suppression system and that he was trained in the use of portable extinguishers during his services in the United States Military. Mr. Doe states in his deposition that he recalls the fire was around 11:00 am. He was working on the char broiler grill. A porta bella mushroom sandwich was being prepared. He testified that there was a normal flash of flame from the balsamic vinaigrette marinade being used on the porta bella mushroom. The flame was no more than a foot and a half high and then went down, lasting no more than a couple seconds. When he turned the mushroom over there was a second flame which was bigger and longer than the foot and a half flame. This flame reached the screens (mesh filters). It hit the screens and he saw it "possibly" go through the screens.

Mr. Doe described seeing the second flame and seeing a huge orange glow on the screen. He further testifies that "It all happened so fast" and that he removed the porta bella mushroom from the grill, saw the orange glow getting bigger in the filter area, assumed the filter was on fire and proceeded to tear off the screen. He looked up, ran and grabbed a dry chemical fire extinguisher which he used on the fire to no avail.

BASIS AND REASONS FOR OPINIONS

The opinions expressed herein are the opinions that I have developed to date based on my education, experience and the information I have been provided to review, including physical evidence, in connection with this matter. These opinions may be revised if additional information becomes available to me.

1.The origin of this fire was on the natural gas fired char grill located in the southeast corner of the basement kitchen. The cause of this fire had been determined to be accidental ignition of grease and/or vinaigrette sauce during the cooking of porta bella mushrooms on the grill. This normal flare-up extended from the char grill grates upward into the mesh filters, into the plenum area and ignited grease in the ductwork to the exhaust fan.

2.The incipient fire and spread of the fire involved the burning of cooking grease and grease-laden vapors.

3.The fire spread into the plenum area of the hood located behind the rack of mesh grease filters and into the exhaust duct with great speed due to the air movement being created by the exhaust fan located at the end of the ductwork on the roof.

4.The fire that extended into the ducts prevented the suppression efforts of the restaurant employees from being successful.

5.This rapid burning fire had extended into the ducts and then spread out into the structural members of the building, preventing the fire department from quickly stopping the fire despite their immediate response upon notification.

6.On the basis of the documents that I have reviewed in connection with this case as well as the physical evidence, it is my opinion that the Suppression Company failed to properly install, service and inspect the fuse link. The *Manufacturer's "Design, Installation & Maintenance Manual"* states: "For protection of the duct, a detector must be centered either flush with the hood-duct opening or in the duct but not more than 23 feet downstream from where the duct is connected to the "hood". (Exhibit 13) The fuse link was actually located on the underside of the top of the hood and was in front of the duct opening. (Exhibit 14) It is apparent that the fuse link did not fuse and separate during the fire thus allowing for the activation of the suppression system by automatic means. As a result, the fire was able to progress beyond the plenum and duct area and cause substantial damage to the structure.

7.It is my opinion that the Suppression System Company failed to properly install, service and inspect the duct nozzle. The *Manufacturer's Design, Installation & Maintenance Manual*, states: "a duct with a perimeter more than 48" but less than 75" requires two nozzles for proper protection". (Exhibit 15) Only one duct nozzle was installed leaving inadequate protection to the duct run.

8.It is my opinion that the Suppression System Company failed to properly install, service and/or inspect the duct nozzle. The *Manufacturer's Design, Installation & Maintenance Manual* states: "The tip of the upper nozzle, of the pair of nozzles required for each duct, shall be positioned in the center of the duct opening and above the plane of the hood-duct opening between 1" and 24". (Exhibit 16) The nozzle protecting the duct was in reality located attached to the top of the underside of the hood and outside of the hood/duct opening. The nozzle was also blocked by the top ridge of the duct opening. (Exhibit 17)

9.It is my opinion that the Suppression System Company failed to properly install, service and inspect the manufacturer's wet chemical system by not sealing the holes left from the removal of the dry chemical system and not causing the holes made for the installation of the wet chemical system to be "liquid tight" in direct violation of *National Fire Protection Association Standard 17, 17A and 96.* (Exhibits 17 & 18) The Fire Prevention Code, Section ____ states: "Commercial kitchen exhaust systems shall be installed and maintained in accordance with NFPA 96". (See Exhibit 19) As a result of this violation of code and standard the fire was able to progress out of the existing holes and ignite the wood structure of the building.

10. It is my opinion that the Suppression System Company failed to properly inspect the manufacturer's wet chemical fire suppression system by not tagging the system as non-compliant. The *State Fire Prevention Code, Section _____*, states: "Any engineered, pre-engi-

neered and self-service fire suppression system inspected and found to be in noncompliance with its listing or manufacturer's specification shall have a service tag attached indicating noncompliance". (Exhibit 20) Had the system been tagged as non compliant, the owner of the restaurant would have had the opportunity to have the violations corrected or the authority having jurisdiction would have been advised of the violations and could have ordered the violations to be corrected before this fire occurred.

11.It is my opinion that the Suppression System Company failed to properly advise its client of the requirements of a "Class K" wet chemical fire extinguisher. *The State Fire Prevention Code, Section* __________, states: "portable fire extinguishers shall be installed and maintained in accordance with NFPA 10". (Exhibit 21) NFPA 10, 1998, Edition, Section 3-7.1, states: "Fire extinguishers shall be provided for hazards where there is a potential for fires involving combustible cooking media (vegetable or animal oils and fats)". (Exhibit 22) Had the proper extinguishers been placed in the kitchen area the suppression effort made with hand held extinguishers by employees may have been successful or at least hampered the progression of the fire until the arrival of the fire department.

12.It is my opinion that the Suppression System Company failed to advise the restaurant owner that mesh filters (Exhibit 23) are not listed for commercial grease vapor removal and that mesh filters are in direct violation of

NFPA 96; and, the manufacturer's listing for the suppression system, NFPA 96, 1998 Edition, Section 3-1, states: "Mesh filters shall not be used". (Exhibit 24) Mesh filters allow grease to be absorbed into the filter causing a "fire load" or a fuel for the fire to consume and to sustain the fire's growth and spread. In this case, the mesh filters increased the fuel load by absorbing cooking grease in the material of the filter and not allowing the heaving grease molecules to be dropped out of the air current and collected into a collection cup.

13.It is my opinion that the Ventilation Cleaning Company failed to advise the restaurant owner that the mesh filters (Exhibit 23) are not listed for commercial grease vapor removal and that mesh filters are in direct violation of NFPA 96 and the manufacturer's listing for the suppression system as set forth in #12, above. In the deposition of the ventilation cleaning company's employee, he admits that the problem with mesh filters is that they catch fire because they accumulate grease and because of the way they are made. (Exhibit 31).

14.In my opinion the Suppression System Company failed to advise the restaurant owner of the violation of NFPA 96 as it relates to the clearance of combustibles to the duct run. NFPA 96, Section 1-3.2.1 states: "ducts shall have clearance of at least 18 in. to combustible material". (Exhibit 25) This improper clearance to combustibles allowed the wood structure of the building to ignite around the duct run.

15.It is my opinion that the Ventilation Cleaning Company failed to advise the restaurant owner of the violation of NFPA 96 as it relates to the clearance of combustibles to the duct run, as set forth and for the reasons stated in #14 above.

16.It is my opinion that the Ventilation Cleaning Company, failed to clean the exhaust duct down to bare metal as required by NFPA 96. As stated in the deposition transcript of Mr. John Doe, the Ventilation Cleaning Company's employee in charge of the cleaning of the Restaurant, that Mr. Dow would clean down to base metal "whenever he could" but the stacks going up to the attic could not be cleaned in accordance with code requirements. Also, Mr. Doe in his deposition (Exhibit Doe-2) circled three areas where he could not get down to bare metal because these areas were hard to reach. (Exhibit 27) Furthermore, the President of the Ventilation Cleaning Company testified in his deposition that NFPA 96 required the kitchen exhaust system to be cleaned down to "bare metal". He defines the term "kitchen exhaust system" as the hoods, duct work and the exhaust fan - in his words "the entire system". It also includes the filters (Exhibit 28 and 29).

17.It is my opinion that the Ventilation Cleaning Company failed to properly secure the access panels in the duct work as required by NFPA 96. Mr. Michael Doe in his deposition admits that he would tape the access panel in the liquor room with duct tape. Mr. Doe also states that there was no caulking around the access panel and that grease could drip out.

He never talked to the Restaurant owner about it and could not think of a reason why he did not do so. (Exhibit 30).

18.It is my opinion that the Suppression System Company improperly certified that the suppression system was in compliance with presently adopted Codes and the manufacturer's listing.

The *Manufacturer's Design, Installation & Maintenance Manual* under Section A-1, states: "The manufacturer's fire suppression system is to be installed, inspected and maintained in accordance with NFPA 17A, *Standard for Wet Chemical Extinguishing Systems.* NFPA 96, Vapor Removal from Cooking Equipment and this manual". (Exhibit 26)

Had the suppression system been installed, serviced and inspected in accordance with the State Fire Prevention Code, National Fire Protection Association Standards, 17A & 96, and the manufacturer's listing and installation and maintenance manual, the fire that occurred at the Restaurant would have been extinguished in the plenum and duct area of the ventilation system without additional damage to the building.

The opinions contained in this report are based on personal observations at the fire scene as well as statements of witnesses as of the date of this report and were developed to a reasonable degree of certainty. The author agrees to a reconsideration of the conclusions if new evidence becomes available and provided.

CASE STUDY II

DISCUSSION

On ____________________, a fire developed on the cooking surface of a gas fired, table top, char grill. The incipient fire spread into and out of the hood ventilation system ad ignited the combustible building materials located between the ceiling and roof. The fire's progression then became horizontal advancing along the roof line, causing the total destruction of the building.

The building was built between 1979 and 1980. At the time of construction two separate ventilation hoods were installed, one above a gas fired, table top, char grill and one over two deep fat fryers. The ventilation hood above the char grill is identified as a manufacturer's ventilation hood. The char grill was protected from fire by a manufacturer's two and a half (2.5) quart wet chemical fire suppression system that had four appliance nozzle drops and <u>did not have any nozzles protecting the plenum and/or duct areas of the ventilation system</u>. The fryer hood had a four gallon wet chemical suppression system. This four (4) gallon wet chemical suppression system had a nozzle over each of the two fryers and nozzles protected the plenum and duct area of the ventilation system. The suppression systems has been installed by a Suppression Systems Company. It is believed that prior to 1997 the systems had been inspected, serviced and maintained by other companies. Since 1997 the suppression systems have been inspected, serviced and maintained by this Company.

The manufacturer's Ventilation Hood is "Listed" and designed to cause grease molecules to be thrown out of the exhaust air stream by cyclonic motion with compression and expansion of the air stream. A cold water mist is continuously sprayed while the ventilation system is in operation. This cold water spray reduced the ambient temperature of the exhaust gases and increases grease molecule extraction. The manufacturer's ventilation system also washes out all grease and residue with a hot water and detergent cleaning cycle activated at the end of the cooking period.(Exhibit 5) The manufacturer's hood system was never "Listed" nor was it intended to act as a fire suppression device proving protection for the plenum and duct.

III REVIEW OF N.F.P.A. 96

A review of *National Fire Protection Association, Standard 96*, "Ventilation Control and Fire Protection of Commercial Cooking Operations", 1998 Edition, Chapter 7, Section 7-2.3, states: "Grease removal devices, hood exhaust plenums, and exhaust ducts requiring protection in accordance with 7-1.1 shall be permitted to be protected by a listed fixed baffle hood containing a constant or fire actuated water-wash system that is listed to extinguish a fire in the grease removal devices, hood exhaust plenums, and exhaust ducts. Each such area not provided with a listed water-wash extinguishing system shall be provided with an appropriate fire suppression system listed for the purpose." (Exhibit 6)

The language of the 1998 Edition is consistent with and clarified the intent of the Standard Committee, including the 1978 standard, which was in effect at the time of this installation. The term "Listed" in Chapter 7, clearly refers to a system that has been tested and listed as a fire extinguishing device and not simply "Listed" as a grease extractor.

REVIEW OF U.L. LETTER

Prior to the date of this fire, Underwriters Laboratories Incorporated was contacted by the undersigned due to numerous investigations that have been conducted in regards to NFPA 96, Chapter 7, Section 7-2.3. Exhibit 7 hereto is a formal letter from UL in response to inquiries made regarding issues pertinent to this case. On Page Two of the letter, the following is stated: "In reviewing our records, we have determined that there are no water-wash type hood and duct systems Listed for the extinguishment of fires occurring in grease removal equipment. As such, the evaluation of the compatibility of the Listed water-wash type hood and duct system for use with Listed wet or dry chemical extinguishing system unit has not occurred. Additionally, of the currently Listed extinguishing system units intended for the protection of restaurant cooking areas, none reference this type of protection in their *Operation, Installation, Inspection, Maintenance and Recharge Manuals.*"

BASIS AND REASONS FOR OPINIONS

The opinions expressed herein are the opinions that I have developed to date based on my education, experience and information that I have reviewed in connection with this matter. These opinions may be revised if additional information becomes available to me.

1.The initial development of the fire was on the tabletop char grill.

2.The incipient fire and spread of the fire involved the burning of cooking grease and grease-laden vapors.

3.The fire spread into the plenum area of the hood and exhaust duct with great speed due to the air movement being created by the exhaust fan located at the end of the ductwork on the roof.

4.The fire that extended into the ducts prevented the suppression efforts of the restaurant employees from being successful.

5.The rapid burning fire that had extended into the ducts then spread out into the structural members of the building which also prevented the fire department from quickly stopping the fire despite their immediate response upon notification.

6.On the basis of the documents that I have reviewed in connection with this case and the physical evidence, the suppression system company failed to properly install the manufacturer's fire suppression system in accordance with manufacturer's instructions and specifications, national and local codes and standards.

7.In my opinion the suppression company failed to properly install the manufacturer's fire suppression system by not properly protecting the plenum and duct areas of the hood with nozzles from the wet chemical extinguishing system. This omission of the nozzles is in direct violation of the UL Listing of the suppression system.

In my opinion the companies performing semi-annual service on the system failed to properly test, maintain, service, replace, repair and inspect the fire suppression system in accordance with its Listing. These companies should have identified the improper installation of the Manufacturer's Suppression System, most notably, the absence of nozzles in the plenum and duct, and reported this condition to the owner or representative as well as make recommendations to correct these deficiencies.

Without proper nozzles located to protect the plenum and duct areas of the hoods, the fire was allowed to advance unimpeded. The existing system did not extinguish the ensuing fire. Upon automatic activation of the system there was not enough nozzle locations for adequate protection. This would not allow for the proper amount of chemicals to be discharged into the protected areas of the ventilation system. It would also not allow for the saponification of the fuel load thereby preventing the ignition of said fuel load to occur.

The opinions contained in this report are based on personal observations at the fire scene as well as the statements of witnesses as of the date on this report and were developed to a reasonable degree of certainty. The author agrees to a reconsideration of the conclusion if new evidence becomes available and provided.

CASE STUDY III

DISCUSSION

As stated in the Fire Department report, the equipment involved in this fire was a "Grease Hood, Duct". (Exhibit 5). This report will address the ventilation and suppression system of the cooking line. Origin and Causation is referred to in the report of Mr. John Does of John Does Associates. The electrical analysis of the appliance of origin is referred to in the report of John Does, P.E.
The appliance line that was under the ventilation hood consisted of from left to right, a 2′6″ wide refrigerator, 36″ wide six burner store, 3′2″ wide double oven, 31″ wide double fryer, 60″ flat top griddle, 28″ wide upright, and two drawer broilers. (Exhibit 6) This cooking line was under a 20′30″ long by 4′6″ wide stainless steel ventilation hood.(Exhibit 7) The ventilation hood was connected to two exhaust ducts each measuring 21.5″ by 11.25″. (Exhibit 8) The exhaust ducts were jointed together via a horizontal transition duct run which ran from west to east above the hood. The horizontal duct then made a ninety degree turn in a southern direction and in the area in front of the east side of the hood changed

direction again and became vertical and traveled in the interior of the building two stories to the roof.

The ventilation hood, exhaust ducts and cooking line were protected by a six gallon wet chemical fire suppression system. The suppression system had appliance nozzles protecting the fryers, flat top griddle and one nozzle located above the oven. (Exhibit 9) The plenum of the hood was protected with five nozzles and the ducts were protected with one nozzle each. (Exhibit 10) Natural gas shut off was done by way of a mechanical gas valve located in the ceiling area of the duct run. (Exhibit 11, photograph V-40). Electric shut down was completed via a micro switch inside the control head. Manual activation of the suppression system was done via a mechanical pull station and automatic activation was done via four, five hundred degree fusible links located against the back wall of the plenum area of the hood. (See Exhibit 12)

The suppression system was being serviced and inspected by ______________ Fire Protection and was last inspected on ___________ with the only noncompliance notice being that the "Upright Broiler piping does not protect the new upright broiler properly and needs to be re-piped." (Exhibit 13) This inspection would have been done in accordance with the *Manufacturer's Design, Installation, Recharge and Maintenance Manual for the Fire Suppression System*, 1982 Edition, Section noted, "Semi-Annual Maintenance Examination" (Exhibit 14); and, *National Fire Protection Association Standard* 96 and 17 A, as listed as part of the systems UL Listing.

The ventilation exhaust system was cleaned by John Doe on ______________________ as noted on their invoice statement, "Hood and Duct cleaning, $340.00". (Exhibit 15) This hood and duct cleaning would have been accomplished using National Fire Protection Association, Standard 96 as a guide.

BASIS AND REASONS FOR OPINION

The opinions expressed herein are the opinions that I have developed to date based on my education, experience and information I have reviewed in connection with this matter. These opinions may be revised if additional information becomes available to me.

1.The incipient fire and spread of the fire involved the burning of cooking grease and grease laden vapors.

2.The fire spread into the plenum area of the hood located behind the rack of aluminum grease filters and into the main exhaust ducts.

3.In my opinion the Fire Protection Company failed to install and/or maintain the proper location for the fuse links. The *Manufacturer's Design, Installation, Recharge and Maintenance Manual for the Fire Suppression System*, 1982 Edition, Page 24, under the Section titled, "Detection Placement Requirements", states: "Exhaust Ducts: Detectors installed in exhaust ducts must be centered at the duct entrance, or at a maximum of 20 ft. (6.1m) into the duct opening". (Figure 46 The fuse link and its housing were not located in the duct opening and in fact were located along the back wall of the plenum area of the hood. This direct violation of the *Manufacturer's Design, Installation, Recharge and Maintenance Manual for the Fire Suppression System,* 1982 Edition, would allow for a delay in the activation of the suppression system which in turn would allow for a longer burn time in the duct area of the ventilation system. This contributed to the progression of the fire outside the protected area of the Suppression System.

4. In my opinion the Fire Protection Company did not inspect the suppression system in accordance with the *Manufacturer's Design, Installation, Recharge and Maintenance Manual for the Fire Suppression System,* 1982 Edition, Section titled, "Semi-Annual Maintenance Examination", in particular under number 12m which states: "Check all nozzles to ensure that they are free of grease build-up and that blow-off caps are properly in place." (Exhibit 14) For clarification of this issue a phone call was placed to Manufacturer's Technical Support and a discussion was held with a representative who stated: "To ensure that the nozzles are free of cooking grease buildup, the nozzles must be taken apart and inspected." The representative further stated: "All manufacturer's distributors are trained to take the nozzles apart to check and ensure that it [the nozzle] is grease free."(Exhibits 16 and 17).

Not only was grease clogging all of the nozzles but also roach infestation and nesting materials were found in some nozzles. This would stop the proper amount of suppressant agent to expel out of the nozzle to extinguish the ensuing fire in any of the protection zones of the system.

5.It is my opinion that the Fire Protection Company did not inspect the suppression system in accordance with the *Manufacturer's Design, Installation, Recharge and Maintenance Manuel for the Fire Suppression System,* 1982 Edition. The mechanical gas valve was not checked for operation. The mechanical gas valve should have been tripped manually during the previous inspection of the system. The photographic and physical evidence indicated that the gas valve did not trip during the fire. During the examination of the gas valve after the fire, said valve was in the set position as was the actuator for the valve in the control head. (Exhibit 11)

6.It is my opinion that the Fire Protection Company's employee failed to clean the ventilation system in accordance with its invoice and National Fire Protection Association Standard 96, "Ventilation Control and Fire Protection of Commercial Cooking Operations, 1998 Edition, Section 8-3, "Cleaning", states under 8-3.1: "Hoods, grease removal devices, fans, ducts, and other appurtenances shall be cleaned to bare metal at frequent intervals prior to surfaces becoming heavily contaminated with grease or oily sludge."

Heavy accumulation of grease allowed for a fuel source of the fire within the ductwork. Due to this heavy fire load of grease the tetrahedron required for open flame was allowed to continue unimpeded until being extinguished by the fire department. This is evident by the high temperature oxidation or patterns on the hood, back wall of the hood and the ductwork.

National Fire Protection Association, Standard 96,*"Ventilation Control and Fire Protection of Commercial Cooking Operations", 1998 Edition, Chapter 4, "Exhaust Duct Systems"*, Section 4-3.1, states: "Openings shall be provided at the sides or at the top of the duct, whichever is more accessible, and at changes of direction". (Exhibit 18). Further, Section 4-3.4.3 states: "On vertical ductwork where personnel entry is not possible adequate access for cleaning shall be provided on each floor."(Exhibit 19)

Had the suppression system for the ventilation system and the cooking equipment been properly maintained, serviced and inspected in accordance with National Fire Protection Association Standard 96, this fire as it progressed into the protected area of the suppression system would have been extinguished before any additional damage could have occurred by the fire's progression out of the ductwork and into the structural members of the building.

The opinions contained in this report are based on personal observations at the fire scene as well as the statements of witnesses as of the date on this report and were developed to a reasonable degree of certainty, The author agrees to a reconsideration if new evidence becomes available and provided.

CASE STUDY IV

DISCUSSION

On June 14, 2002, and again on June 18, 2002, the undersigned examined, documented and photographed the loss location. Evidence was retrieved, identified and secured to storage. (See exhibit 5). During the examination of the loss site on_____, a discussion was held with the origin and cause investigator, with the following information obtained:

During the morning of _____, at approximately 0900 hours, a fire occurred in a salamander located below a tabletop char grill. The ensuing fire progressed into the plenum area of the ventilation hood (area behind the grease filters) and into the exhaust ducts of the ventilation system. The fire then extended outside the confines of the ventilation system and into the "class A" structure of the building. A review of the deposition testimony of Mr. _____, the executive chef of the restaurant, confirms that the progression of the fire (See exhibit 6).

A review of the deposition transcript of _____, a prep cook at the restaurant, indicates that he (____) began to cook at 0700 hours the day of the loss. After cooking chicken breast, flank steaks, eggplants and portabella mushrooms for approximately two hours, he removed the items, left the grill on and proceeded down to the basement area to assist with a delivery that had arrived. He heard someone yell fire from the parking area. Mr. ____ proceeded back up the stairs to the kitchen. When coming up to the kitchen, Mr. _____ saw no fire but he saw the filters melted out over the char grill. Mr. ____ opened the kitchen egress door and looked outside to the area of the exhaust duct-work. The exhaust duct was "fully involved in fire". Mr.____ then proceeded to use an extinguisher and also tried the manual pull of the suppression system to no avail (See exhibit 7).

During the examination of the loss site, the kitchen was noted to be located in the middle of the south wall of the building. The cooking line was located in the southeast corner of the kitchen. The cooking line, from left to right or east to west, consisted of a 34 x 36″ table top char grill with a single tray salamander located underneath, a 16″ x 32″ fryer and a ten-burner 60″ x 30″ stove with two ovens below the burners (See exhibit 8). The cooking line was located under a 10′ x 53″ ventilation hood that had two 17″ x 17″ duct openings (See exhibit 9). The suppression system was identified as a Manufacturer's dry chemical fire suppression system. The suppression system had six appliance nozzles, two plenum nozzles and two duct nozzles (See exhibit 10). Automatic activation of the suppression system was accomplished by mechanical means by the utilization of three metal alloy fusible links (See exhibit 9). John Doe Suppression Company, located at _______, last serviced the suppression system on ____ (See exhibit 11).

John Doe, Incorporated, located at _____, last cleaned the ventilation system of the commercial cooking line on ________ (See exhibit 12).

III. BASES AND REASONS FOR OPINIONS

The opinions expressed herein are the opinions that I have developed to date based on my education, experience and information I have reviewed in connection with this matter. These opinions may be revised if additional information becomes available to me.

1. The initial development of the fire was within the salamander located under the char grill of the cooking line. The incipient fire and spread of the fire involved the burning of cooking grease and grease-laden vapors. The fire spread into the plenum area of the hood, located behind the rack of aluminum grease filters, and into the main exhaust duct with great speed due to the air movement being created by the exhaust fan located at the end of the ductwork on the roof. The fire that extended into the ducts prevented the suppression efforts of the restaurant employees from being successful. This rapid burning fire that had extended into the ducts then spread out into the structural members of the building, which also prevented the fire department from quickly stopping the fire despite their immediate response upon notification.

2. On the basis of documents that I have reviewed in connection with this case and the physical evidence, Joe Doe Suppression Company, failed to install, service and maintain the required amount of fusible links in the venti-

lation hood. The Manufacturer's System Instruction Manual, Page 9-3 states:

"Detector location"

A fusible link must be located above each cooking appliance (in the exhaust air flow) and in each duct.

If the duct is located directly above an appliance the fusible link above that appliance can be deleted. The link should be located within the hood area above, or below, the bank of filters. Figure 9-5 shows the location of the fusible links, with respect to the filters." (See exhibit 13)

None of the three existing fuse links within the plenum area of the hood were located over or in the exhaust airflow. This violation of the UL Listing of the suppression system would cause a delay in activation. This delay would allow the fire to grow within the ventilation system and impinge upon the combustible structure of the building.

3. On the basis of documents that I have reviewed in connection with this case and the physical evidence, John Doe, failed to properly advise their client of the improper or unbalanced piping of the system. The Manufacturer's Instruction Manual, Page 4-6 states:

"Piping unbalance"

The piping may be unbalanced on either side of any tee. The total piping length, on one side

of T1, must not be more than three (3) times that of the other side. (See Figure 4-5)." (See exhibit 14)

The east appliance run that protected the char grill had a ratio of approximately four (4) times the other side. One side of the piping was 6" and the other 22". This would **not allow** the extinguishing chemical to be properly or equally proportioned from each nozzle. Without the proper amount of agent being expelled from each nozzle of the system, non-extinguishment of the fire would occur (See exhibit 10).

Mr. Joe Doe, president of John Doe Sales and Service LLC, in his deposition when asked:

"If the system is not balanced what can happen?"

Answered: **"It will discharge more out of one nozzle than the other nozzle."** (See exhibit 22)

It is obvious that Mr. John Doe understands the significance of an unbalanced system but fails to understand the ramifications. This is clear in Mr. John Doe answer to the following question during his deposition:

Question: **"Do you use noncompliant tags?"**

Answer: **"When I get a noncompliant system I do"**

Question: **"What's it mean to have a noncompliant system?"**

Answer: **"It means there's a major deficiency**

that won't make the system operate." (See exhibit 22)

4. On the basis of documents that I have reviewed in connection with this case and the physical evidence, John Doe Suppression Company, failed to properly inform their clients of the non-protection of the salamander by the suppression system. The Manufacturer's Instruction Manual, Page 2-2 states:

"Each piece of kitchen equipment must be protected by the suppression system described in Section 1." (See exhibit 15)

Page 3-3 of the Manufacturer's Instruction Manual states:

"BROILER NOZZLE PLACEMENT AND COVERAGE

The upright broiler is protected by a B-1 nozzle located beneath the grate and above the grease deflector plate." (See exhibit 16)

Again Mr. John Doe misunderstanding of the Manufacturer's manual is evident by his own deposition testimony, as the following shows.

Question: **"Is the grease tray-do you see the grease tray depicted in these two photographs?"**

Answer: **"Yes"**

Question: **"Is that something that needs separate protection?"**

Answers: **"No"**

Question: **"Why not?"**

Answer: **"I guess you'll have to ask the people who wrote the code. The code doesn't require that being protected."** (See exhibit 23)

Mr. John Doe's interpretation of the Manufacturer's Manual is wrong. Mr. John Doe lack of qualification to interpret the Manufacturer's Manual, a manual that was written and taught to the Manufacturer's distributors 30 years before Mr. John Doe even entered the kitchen suppression system inspection business, is again blatantly evident.

5. On the basis of documents that I have reviewed in connection with this case and the physical evidence, John Doe Suppression System Company, failed to properly inform their clients of the requirement for a Class K extinguisher to supplement the dry chemical suppression system. NFPA 10, "Standard for Portable Fire Extinguishers" section 4.3.2* states:

"Class K Fire Extinguishers for Cooking Oil Fires"

Fire extinguishers provided for the protection of cooking appliances that use combustible cooking media (vegetable or animal oils and fats) shall be listed and labeled for Class K fires." (See exhibit 18)

6. On the basis of documents that I have

reviewed in connection with this case and the physical evidence, John Doe Suppression Company failed to warn their client of the non compliance of the ventilation system from combustible materials. NFPA 96, "Ventilation Control and Fire Protection of Commercial Cooking Operations" Section 4.2.1 states:

"Where enclosures are not required, hoods, grease removal devices, exhaust fans, and ducts shall have a clearance of at least 457mm (18in.) to combustible material, 76mm (3in.) to limited-combustible material, and 0mm (0in.) to noncombustible material." (See exhibit 17)

The UL Listing of the system under test protocol requires the suppression system to be installed within a compliant NFPA 96 ventilation system.

The ventilation hood and duct system was directly against the combustible material of the building. This violation allowed for the fire to ignite the exterior wall of the building instead of being encapsulated within the confines of the ventilation system.

Mr. John Doe agrees in his own deposition that he has a duty to advise his client of the violation to combustibles. When asked:

Question: **And at the bottom there it says, on this date the above system was tested and inspected in accordance with the procedures from the presently adopted additions of NFPA 17, and 17-AA (sic), 96 and the manufacture's**

manual with the results indicated above. What does that mean?"

Answer: **"That means I've checked out this system according to those three standards and anything that's unique to that system according to the manufacture."** (See exhibit 24)

It is again obvious that Mr. John Doe violated his own inspection requirements when the inspection of the suppression system was done.

7. On the basis of documents that I have reviewed in connection with this case and the physical evidence, Mary Jane, Incorporated failed to properly clean the ventilation system in accordance with NFPA 96, "Ventilation Control and Fire Protection of Commercial Cooking Operations" Section 11.4.2*, which states:

"Hoods, grease removal devices, fans, ducts, and other appurtenances shall be cleaned to bare metal prior to surfaces becoming heavily contaminated with grease or oily sludge." (See exhibit 19)

Mary Jane, Incorporated indicates on its inspection tag that non-accessible areas of the ventilation system exist. Based upon an examination of the entire ventilation system there is no, non-accessible areas and all areas of the ventilation system could have been cleaned to bare metal.

Mary Jane, Inc. failed in their responsibility to their client to properly clean all areas of the ventilation system, thereby allowing an accumulation of grease to become a "fire load" hazard.

Further, John Doe Suppression Systems LLC failed in their professional responsibility to their clients, ABC, Inc. t/a a Restaurant, by improperly inspecting the Manufacturer's suppression system involved in this loss, in violation of Manufacturer's Design, Installation and Maintenance Manual and National Fire Protection Association Standards 17 and 96.

Even though John Doe LLC noted the piping unbalanced and the requirement of the Class K extinguisher, they (John Doe) failed to advise their client of the meaning of the notations on the inspection form of the suppression system.

Mr. Restaurant, owner of the restaurant, when asked during his deposition:

Q "What are the in specifics that you remember? The lack of specifics?"

Testified:

A. "I remember getting an invoice from John Doe with a bunch of items on it that I didn't understand specifically. And I called him and asked them what they meant. And he told me that there were something's that could be changed in the system. And I asked him if our system would work the way that it was and he said yes. And

I said is our system certified the way it is? And he said yes." (See exhibit 20)

Fire Marshall _____, when asked in his deposition:

Q. "Do you as a Code Enforcement Official when you are inspecting a restaurant personally rely on this tag?

Testified:

A. "Yes, yes absolutely. If their tag isn't there, we require them to have a company come in to do the inspection and to make sure that they secure the proper tag on the system." (See exhibit 21)

Both the owner of the restaurant and the Code Enforcement Official relied on the professionalism of John Doe and Mary Jane to properly and professionally ensure that the ventilation system and suppression system were in accordance with code and standard.

The opinions contained in this report are based on personal observations at the loss scene; statements of witnesses and other documents reviewed as of the date on this report and were developed to a reasonable degree of certainty. The author agrees to a reconsideration of the conclusion if new evidence becomes available. This investigation was conducted using NFPA 921 as a guide and other authoritative sources.

CASE STUDY V

DISCUSSION

Based upon various documents reviewed (See exhibit 4), the causation of the fire was a rag left on the flat top griddle by John Doe Suppression Company technicians, ____ and ____, who approximately three hours before the fire had been in the kitchen performing a six month inspection of the Fire Suppression System which protected the ventilation system and cooking line against fire.

As stated in the report of ____, Deputy State Fire Marshal, *"On ______, at approximately 0308 hours a fire was reported to the _a_ Dispatch Center that a fire was in progress at the a Hotel Casino. Fire crews from the City of __ Fire Department responded at that time. Upon arrival the fire crews went into the kitchen area and found heavy smoke conditions and flames coming from between two cooking grills. The fire had extended into the confined spaces above the kitchen hood system. Fire Chief ____ notified the ___ City office of the fire at approximately 11:00 am on ______, informing Acting Chief Deputy _____ of the incident and that the fire suppression system in the hood and duct had failed to discharge during the fire."*

"On _______, the Fire Marshals Office conducted a meeting with interested parties. During this meeting videotape from a security camera was reviewed which showed two individuals working on the suppression system on the night

of the fire. The examination included the bagging and securing of the systems discharge nozzles to contain the system chemical in the event of a discharge, the examination of the firing mechanism which was found in the cocked position." (See exhibit 5 & 6)

REVIEW OF MANUFACTURER'S, DESIGN, INSTALLATION, MAINTENANCE & RECHARGE MANUAL

The manual, under section noted as, "MECHANICAL RELEASE MODULE" states;

"The Mechanical Release Module is used to actuate the Agent Cylinder/Discharge Valve either automatically or manually by puncturing a Nitrogen cylinder. The pressure from the cylinder pneumatically opens the discharge valve(s).

Automatic Release of agent is accomplished when a Fusible Link Detector separates under a fire condition and release tension on the cable. This causes a spring-loaded plunger to perforate the cylinder seal and releases nitrogen through the Actuation Hose/ Piping Network to the Discharge Valve(s).

Manual release of agent is accomplished by pulling on a Manual Pull Station, which is connected, to the Mechanical Release Module by a cable."

In actuality there is a mechanical step missing in the above statement. When a fuse link separates and tension on the cable is released the cable that is connected within the control head to a plate, noted as the "link plate", this plate rotates in a counter clockwise rotation due to a spring (tension bar) that begins to compress after cable tension is released. A "link plate cam" that is attached to the back of the link plate (near the twelve o'clock position) rotates with the link plate, striking the "activation scissors", causing the scissors that are hinged to pivot on its center axis. When this occurs the puncturing pin activates the nitrogen cartridge.

The manual further states that when attaching the cable to this cam plate that the, Special Tool, over the manual pull cam housing until it rests against the outside edge of the link plate. Draw tension on to the cable through the connector until the link plate is drawn against the set-up tool, then tighten the set screw on the connector." (See exhibit 7)

In other words, the Safety Tool, allows the link plate to be set is the proper location. The second tool that is utilized known as the cocking tool allows the hinged scissors to be set in a straight vertical position. This tool attaches to a ratchet drive with an extension drive attached.

DEPOSITION OF SYSTEM SERVICE TECHNICIAN

In the deposition transcript of ________ the following facts are stated:

Q. "**So are you basically testing this thing to**

fire both the manual station first and then the automatic version?"

A. **"YES"**

Q. **"Okay. When you pulled the manual station, the system fired up link it should have?"**

A. **"Yes"**

Q. **"And when you cut the terminal link, did it fire like it should?"**

A. **"I'm not sure if it did. I think that's where the problems started coming up." (See exhibit 8)**

Mr. _____ developed the hypothesis that the problem was grease within the conduit line of the link detection or grease on the link scissors that hold the fuse link in place.

Q. **"What did you do in response to that?"**

A. **"We knew that we had to get the link holders out of there to get the line free because with this grease, certain type of grease is set up they turn into like to silicon gel and it will prevent anything from moving. So even if you cut the link things may not fire. So the object was to free the line so we had to get those link holders off of there, make sure the line was clean and then put the links on with new holders."** (See exhibit 9)

Q. **"The detection bracket and the link holder that you replaced, was that replaced with the**

Manufacturer's part?"

A. **"They were another manufacturer's parts."**

Q. **"Was there a problem using ABC part in XYZ system?"**

A. **"No."** (See exhibit 10)

After the brackets were replaced the system was not tested.

Q. **"After you have replaced the link holders did you test the detection line at all by cutting the link to see if it had fixed the problem you identified earlier?"**

A. **"No."** (See exhibit 11)

The second hypothesis that the service technician developed while trying to re-cock the system was that the new link holders were not allowing for the proper amount of cable length.

Q. **"After you re-cocked the system, do you put the nitrogen actuator- - -**

A. **"I was having problems re-cocking the system because the line now because of the change in the link holders, it was a little tighter. So I couldn't put the new _ there's a little barrel or pin that goes in one of the slots for detection line so that when you go to cock it has to be there and you push in that little scissor mechanism to cock it and it wasn't allowing me to cock it.** (See exhibit 11)

Mr. ___ was not able to simply readjust the cable slack due to the fact that the cable adjustment screw had previously broken off.

A. **"Got it. The barrel was in there and one thing we noticed was that there was a screw in there that was broken off but it was still _ the barrel was on there and it was tight. So we just, instead of cutting that off and having to replace the whole thing, we just left it the way it was and tried loosening the line up. I figured it I worked with it a bit it would loosen up to were I could put it in the slot and then I could cock the system." (**See exhibit 12)

Again after Mr. ___ had pulled on the cable enough to stench it, he then was able to cock the system but did not test fire it.

Q. **"So how did you go about getting a little play out of the detection line?"**

A. **"Just kept pulling on it until I got enough slack to were I could put it in the slot and then I was able to cock it."**

Q. **"At that point did you do another dry run or testing of the system to see if it fired?"**

A. **"No."** (See exhibit 13)

Mr. ____ also states that during this inspection of the Manufacturer's system, he did not have the Manufacturer's special tools to properly cock and safety the system. Instead he utilized a large screwdriver.

Q. **"So you never had the __Special__ tool?"**

A. **"No. Just used a big screwdriver."** (See exhibit 14)

Mr. _____ further states that he also did not have the Manufacturer's manual with him.

Q. **"When you were doing your service call at the __Loss Location__, did you bring out the Manufacturer's manual at all?"**

A. **"No."** (See exhibit 15)

Mr. ____ further states that he has never reviewed any inspection procedure manuals by his employer John Doe Suppression Company.

Q. **"John Doe Suppression Company document __XXX__ entitled Components and Operation Kitchen Hoods and Suppression Systems, ever seen this before?"**

A. **"No."**

Q. **"John Doe Suppression Company document __XXX__ entitled Maintenance Procedures Kitchen Hoods Suppression Systems, have you ever seen that before?"**

A. **"No."** (See exhibit 16)

After the fire when Mr. ___ inspected the (MCH) mechanical control head, he noticed that the fuse link line was slack but the scissors were still cocked.

Q. **And you were able to see the detection line slack?"**

A. **"Yes."**

Q. **"And the links were melted?"**

A. **"Yes."**

Q. **"The scissors should pop?"**

A. **"Right."**

Q. **"And they didn't?"**

A. **"Right. Still cocked."** (See exhibit 17)

TESTING OF MANUFACTURER'S CONTROL HEAD

During the course of the investigation into this loss by the undersigned, testing of an identical control head was performed. To recreate the suppression system fuse link line that was in place at the restaurant at the time of the loss various documents were utilized. These documents consisted of, but not limited to the deposition transcripts of the John Doe Suppression Company employees, Fire Marshall report, Manufacturer's Manual, ____ Report, etc. The control head was attached to a fuse link line utilizing the same components as descript in various documents. The system was cocked after Manufacturer's scissors were installed and the system was fired, numerous times from the terminal link. The system was always re-cocked and safety utilizing the proper Special tools. The scissors were then changed out to other manufacturer's components and the system was again cocked and fired. After numerous cocking and resetting of the control head, the system was fired and the link plate cam snapped off from the rear of the link plate. The link plate cam was retrieved, identified and secured as evidence.

After the link plate cam snapped off during normal tripping and resetting of the control head, the system was cocked and fired by cutting of the terminal link. The link plate rotated in a counter clockwise position but with the missing cam not being in its proper place the scissors did not trip. It must be noted that the same problem with the cable not being able to be placed in its slot on the link plate was also noticed as descript by Mr. _____ in his deposition. This problem of inadequate link line length was occurring with the Manufacturer and Other Manufacturer's scissors. The problem of the cable length was not caused by the Other Manufacturer's scissors but in fact that the link plate had rotated to its tripped position. Mr. ______ failed to recognize that the link plate was in the tripped position when re-arming the system, which accounted for the difficulties he described in trying to rearm the system. (See exhibit 24, video of testing)

EXAMINATION OF PHYSICAL EVIDENCE

On ______, the undersigned traveled to the

State Fire Marshal Office in a City, State and an examination of the evidence from the fire being retained by the Fire Marshal was inspected. During the examination of the control head it became apparent that the link plate cam was also missing from that control head.

III. BASES AND REASONS FOR OPINIONS

The opinions expressed herein are the opinions that I have developed to date based on my education, experience and information I have reviewed in connection with this matter. These opinions may be revised if additional information becomes available to me.

1. The initial development of the fire was rags left on a flat top griddle of the cooking line. On the basis of documents that I have reviewed in connection with this case and the physical evidence, John Doe Suppression System Company through its employees _____ and _______, breached their standard of care of a reasonably prudent fire suppression system technician by failing to remove the rag or other similar combustible material from the surface of a cooking appliance at the conclusion of their inspection during their final walk-through.

2. The incipient fire and spread of the fire involved the burning of cooking grease and grease-laden vapors.

3. The fire spread into the plenum area of the

hood, located behind the rack of grease filters and into the branch and main exhaust ducts.

4. This rapid burning fire that had extended into the ducts then spread out into the structural members of the building which also prevented the fire department from quickly stopping the fire despite their immediate respond upon notification.

5. On the basis of documents that I have reviewed in connection with this case and the physical evidence John Doe Suppression Company through its employees ______ and ______ failed to properly inspect the Manufacturer's suppression system by not utilizing the proper tools. As stated in the Manufacturer's manual when attaching the cable to this cam plate that the, "*Special Tool, over the manual pull cam housing until it rests against the outside edge of the link plate. Draw tension on to the cable through the connector until the link plate is drawn against the set-up tool, then tighten the set screw on the connector.*" (See exhibit 7)

Without the proper tool being utilized the link plate will not be reset into its cock position. The link plate still under the tension of the cocking spring will remain in the tripped on fired position. This will not allow the fuse link line to be properly set into the link plate. The frayed detection line and broken screw also point to the fact that the inability to reset the link plate in its proper position was a problem of long duration.

The illusion that there is not enough slack in the cable will be indicated, as was exactly what occurred when Mr. ____ tried to place the link line pin into the link plate. Further, the improper use of a screwdriver often causes damage to the link plate or scissors of the suppression system. The failure of John Doe Suppression Company, through its employees _______ and _____, to use the proper tools during their service call was a breach of the standard of care and precluded John Doe Suppression Company from identifying and remedying the break in the link plate cam.

Mr. ____ also states that during this inspection of the Manufacturer's system, he did not have the Special tools to properly cock and safety the system. Instead he utilized a large screwdriver.

Q. **"So you never had the Manufacturer's tool?"**

A. **"No. Just used a big screwdriver."** (See exhibit 14)

6. On the basis of documents that I have reviewed in connection with this case and the physical evidence, John Doe Suppression Company through its employees _____ and _____ failed to properly inspect the Manufacturer's suppression system by not discovering the presence of the broken cam, which rendered the Manufacturer's suppression system non-opera-

tional. National Fire Protection Association, Standard 17A, section 5-3.1.1(f) states, **"Where maintenance of the system(s) reveals defective parts that could cause an impairment or failure of proper operation of the system(s), the affected parts shall be replaced or repaired in accordance with the manufacturer's recommendation."** (See exhibit 21)

The Manufacturer's maintenance manual, Section VII-1, required John Doe Suppression Company to perform a "**complete functional test**" during the ____ service call. A proper and thorough testing of the detection system by John Doe Suppression Company, as was called for by the Manufacturer's maintenance manual, would have disclosed that the system was not firing nor was it capable of firing. (See exhibit 22)

7. On the basis of documents that I have reviewed in connection with this case and the physical evidence John Doe Suppression Company through its employees _____ and _____ failed to properly inspect the Manufacturer's suppression system by not being properly trained on the Manufacturer's suppression system, in violation of the Manufacturer's maintenance manual, Section VII, page VII-I, which states: "Maintenance shall be performed by a factory trained, authorized Manufacturer's restaurant system distributorˇ"
(See exhibit 22)

This improper training is in direct violation of National Fire Protection Association, Standard 17A, "Standard for Wet Chemical Extinguishing Systems" section 7.3 Maintenance, states, **"A trained person who has undergone the instructions necessary to perform the maintenance and recharge service reliability ad has the applicable manufacturer's listed installation and maintenance manual and service bulletins shall service the wet chemical fire-extinguishing system 6 months apart as outlined in 5-3.1.1"** (See exhibit 18)

Mr. ____ in his own deposition transcript states that he has only done approximately six Manufacturer's systems prior to the one involved in this loss, that he did not have the manufacturers manual or bulletins with him when he inspected the suppression system and finally, that he is not factory trained to inspect the systems and does not even know if his employee John Doe Suppression Company is an authorized distributor. This was a violation of National Fire Protection Association Standard 17A, section 5-3.1 and the Amerex maintenance manual, Section VII.

Q. **"When you were doing your service call at the Loss Location, did you bring out the Manufacturer's manual at all?"**

A. **"No."** (See exhibit 15)

Q. **"At the time you were doing this service call at the Loss Location, had you been factory trained technician from the Manufacturer?"**

A. **"No."** (See exhibit 19)

Q. **"Do you know if John Doe Suppression System Company was an authorized Manufacturer's restaurant system distributor?"**

A. **"I don't know"** (See exhibit 20)

8. On the basis of documents that I have reviewed in connection with this case and the physical evidence John Doe Suppression Company through its employees _______ and _____ failed to properly inspect the suppression system by not testing the system after their suppose repair to rectify the firing problem were done.

Again, in his own deposition transcript, Mr. ____ states that after he had pulled on the cable enough to stench it, he then was able to cock the system but did not test fire it.

Q. **"So how did you go about getting a little play out of the detection line?"**

A. **"Just kept pulling on it until I got enough slack to were I could put it in the slot and then I was able to cock it."**

Q. **"At that point did you do another dry run or testing of the system to see if it fired?"**

A. **"No."** (See exhibit 13)

This non-test of the suppression system is in direct violation of the Manufacturer's manual and NFPA 17A which states, under section 5-3.1.1(e): **"All wet chemical systems shall be tested, which shall include the operation of**

the detection system signals and releasing devices, including manual stations and other associated equipment." (See exhibit 21).

This non-test was in direct violation of the Manufacturer's maintenance manual, Section VII, page VII-1, as well, which states, **"A complete functional test of the system as described in the "Testing and Commissioning" section of this manual shall be performed at every maintenance interval..."** The Testing Commissioning section, Section VI, page VI-1, states: **"The TERMINAL DETECTOR must have a test link installed in place of the fusible link. With the actuation cylinder removed from the MRM, the release mechanism cocked, the tension bar in the "up" (tension) position and the Special tool removed, cut the test link with a wire cutter or similar device. The MRM must fire at this time."**

9. On the basis of documents that I have reviewed in connection with this case and the physical evidence John Doe Suppression System Company through its employees _____ and _____ failed to properly inspect the Manufacturer's suppression system by utilizing another manufacturer's scissors or links holders in the detection bracket. The Manufacturer's maintenance manual, Section I, General Information, states that **"these systems shall be installed in accordance with the manufacturer's installation manual (which is referenced as part of its listing) and only system components referenced in this manual are used."** The detector linkage component referenced by the manual

includes only the Manufacturer's Detector, part number ###. Section III. Page III-5. The manual does not reference the use of another manufacturer's parts. (See exhibit 23)

10. On the basis of documents that I have reviewed in connection with this case and the physical evidence John Doe Suppression Company through its employees _____ and _____ failed to comply with and violated XXX 477.315.1 (i)(2) by rendering the Manufacturer's fire suppression system inoperable during the course of their service call on ________.

11. On the basis of documents that I have reviewed in connection with this case and the physical evidence John Doe Suppression Company through its employees ______ and ______ breached their standard of care as reasonably prudent fire suppression system technicians by tagging and certifying the system as operational without testing that system after making repairs and replacing parts of the detector equipment, and tagging a system that was not operational.

12. John Doe Suppression Company failed in their professional responsibility to their clients, _____ Incorporated, by improperly inspecting the Manufacturer's suppression system involved in this loss in violation of the Manufacturer's Design, Installation and Maintenance Manual, National Fire Protection Association Standards 17A and 96.

The opinions contained in this report are based on personal observations at the loss scene; statements of witnesses and other documents reviewed as of the date on this report and were developed to a reasonable degree of certainty. The author agrees to a reconsideration of the conclusion if new evidence becomes available. This investigation was conducted using NFPA 921 as a guide and other authoritative sources.

CASE STUDY VI

DISCUSSION

Based upon the City Fire Marshals report and the deposition transcript of ______, the following events occurred on the day of the fire. Mr. ____ stated that he was working in the kitchen preparing some plates for the server to bring to customers, when he noticed that the pot in which he had oil heating up to fry some beef tenders, had burst into flames. As he watched the pot burning, he also noticed the filters above the pot beginning to melt. He then observed the wet chemical agent coming from a separated pipe at the suppression agent storage cylinder but not from the suppression system nozzles (See exhibit 18). A City police officer then arrived and evacuated the restaurant.

On _______, all interested parties conducted a joint investigation, during which time the loss site was documented and photographed (See exhibit 5). Evidence of interest was retrieved and retained on these dates (See exhibit 6).

An examination of the kitchen revealed two cooking lines; one north and one south cooking line. The fire occurred on the stove of the north cooking line. From left to right (west to east), the north cooking line consisted of a 38″x36″ convection oven, a 5′x32″-10 burner stove, a 21″x40″ fryer and a 4′x38″ flat top griddle (See exhibit 7). The cooking line was positioned under a 12′7″x54″ ventilation hood, which had a 2′ extension added onto the east, right side. The 16″x16″ exhaust duct connected at the top center of the original 12′7″hood and ran vertically through the roof line to an exhaust curb and fan. A Manufacturer's 4 gallon wet chemical suppression system protected the ventilation system and cooking line from fire. The suppression system consisted of six appliance nozzles, two plenum nozzles (area of hood behind grease filters), and two duct nozzles (See exhibit 8).

John Doe Fire Protection installed the original suppression system along with the original ventilation hood and ductwork. John Doe also relocated the Manufacturer's control head and cylinder. The suppression system was last inspected by John Doe Fire Protection on ________. In its report issued to the State Fire Marshal's Office for this inspection, John Doe did not note any violations and stated "**SYSTEM PASS.**" (See exhibit 9).

III. BASES AND REASONS FOR OPINIONS

The opinions expressed herein are the opinions that I have developed to date based on my education, experience and information I have reviewed in connection with this matter. These opinions may be re-vised if additional information becomes available to me.

1. The initial development of the fire was within a pot that was located on the ten-burner stove under the north hood of the main cooking line.

2. The incipient fire and the spread of the fire involved the burning of cooking grease and grease-laden vapors.

3. The fire spread from the pot into the plenum area of the hood, located behind the rack of aluminum grease filters and into the exhaust duct with great speed due to the air movement created by the exhaust fan, located at the end of the ductwork on the roof.

4. This rapid burning fire that had extended into the duct was able to spread to the structural members of the building, because the main supply pipe of the Manufacturer's suppression system separated upon when the system activated. This separation of piping at the union of the main supply pipe prevented the suppression agent from extinguishing the fire within the protected area of the suppression system. In deposition, ________ stated, when asked:

Q. "**At any time that day after the fire started did you see the pipe and the coupling come apart?**"

A. "**Yes. Yes. And, in fact, the liquid that was supposed to come out to combat the fire just spewed out.**

Q. "**I guess the question is, the real question I want the answer to, did you see the pipe come**

out of the coupling physically?"

A. " **I did not see it actually become undone. What I saw was the liquid spewing all over the place."**

Q. **"Did you see any liquid come out of the various nozzles that were on the pipe that's in photograph 2 at any time?"**

A. **"No."** (See exhibit 10)

5. On the basis of documents that I have reviewed in connection with this case and the physical evidence, John Doe Fire Protection Systems Incorporated failed to reinstall the Manufacturer's suppression system by not applying for a permit as required by the Authority Having Jurisdiction (AHJ). _______ in his deposition stated:

Q. **"Did your company apply for any permits to disconnect the tank?"**

A. **"No, we didn't."** (See exhibit11)

Further, in the deposition transcript of City of _____ Fire Marshall _______ stated, when asked:

Q. **"Do you know if any permits were taken out for the fire suppression system in year?"**

A. **"Not that I could find."**

Q. "**If a permit was taken out in <u>year</u> for the extension of the hood within that restaurant, would then the City of ______ Fire Marshal's Office have gone out and done an acceptance test?**"

A. "**Yes. There would have been two permits. There would have been one for the hood extension, because you're modifying the hood, and there would have been one for the fire suppression, since they had to apparently move the tank or if they had to add any drops.**

So there would have been two permits that would be taken out. Plans would have had to been submitted for showing us what they were going to do on the hood and what they were going to do with the upgrade to the fire suppression system.

We would have reviewed those plans and then at that time we would have went out upon a call from both - - either one of those contractors would have gone out and done an acceptance test on the fire suppression system and would have looked at the hood to make sure they were meeting the requirements of NFPA 96." (See exhibit 19)

6. On the basis of documents that I have reviewed in connection with this case and the physical evidence, John Doe Fire Protection Systems Incorporated failed to reinstall the <u>Manufacturer's</u> suppression system by not properly allowing for the length of threaded pipe

to be attached tightly into the union. The Manufacturer's, Design, Installation & Maintenance Manual, Section 3-23 states, **"All pipe and fittings must be made tight without pipe dope or thread sealant."** (See exhibit 12)

_________ in his deposition transcript stated, when asked:

Q. **"If the system had worked properly when it activated, should that piping have separated there?"**

A. **"No."**

Q. **"Do you know why it separated?"**

A. "If it separated at all, it was because it wasn't connected at all."

Q. **"Do you know if your company connected it?"**

A. **"They were supposed to have connected it. It should have been connected."** (See exhibit 13)

7. On the basis of documents that I have reviewed in connection with this case and the physical evidence, John Doe Fire Protection Systems Incorporated failed to properly reinstall the Manufacturer's suppression system by not advising its client that the extension of the hood and addition of larger ap-pliances required the installation of a larger Manufacturer's cylinder. John Doe Fire Protection Sys-tems should have replaced the existing

four-gallon cylinder with a six-gallon cylinder when the hood was extended. ________ stated in his deposition transcript:

A. "**And the cylinder was maxed out. It already was using 12 flow numbers, so you have to go a six-gallon cylinder.**" (See exhibit 14)

8. On the basis of documents that I have reviewed in connection with this case and the physical evidence, John Doe Fire Protcction Systems Incorporated failed to properly install the original ventilation hood by not installing the hood with the proper clearances from combustibles. The hood was only 2" from the combustible wall that it was attached to. NFPA 96 requires 3" to limited combustible materials and 18" to combustible materials. ________ in his deposition transcript agrees with this opinion when asked he states:

Q: "**Does that appear to be properly spaced from combustible materials?**"

A. "**No.**" (See exhibit 15)

Further, had the ventilation system and suppression been installed in compliance with NFPA 17A, NFPA 96 and the manufacturer's listing, the fire would have been extinguished while still in the protective zone of the suppression system. Again, ________ shares and agrees with this opinion. When asked in his deposition, he stated:

Q. "**Let me ask you this. If you have a cooking**

fire on a range and you have a fire suppression system above that range that is compliant with NFPA 17A, NFPA 96 and the manufacturer's in-structions, would you expect that system to put that fire out?"

A. "**Yes.**" (See exhibit 16)

Q. "**It also states down here all work to comply with all state and local codes. Are the codes you are talking about NFPA 96, NFPA 17A, and manufacturer standards?"**

A. "**Yes.**" (See exhibit 17)

The opinions contained in this report are based on personal observations at the loss scene; statements of witnesses and other documents reviewed as of the date on this report and were developed to a reasonable degree of certainty. The author agrees to a reconsideration of the conclusion if new evidence becomes available. This investigation was conducted using NFPA 921 as a guide and other authoritative sources.

CASE STUDY VII

DISCUSSION

As stated above, a fire loss occurred at the above referenced loss site. Based upon documents reviewed, deposition transcripts, the report of the ____ State Fire Marshal and a detailed examination of the loss site, the fire originated at the exhaust vent of a Manufacturer's, three vat gas fryer during the morning prep time. A detailed analysis of the origin and cause of the loss is addressed in the report of _____, CFI. This report will discuss the non-extinguishment of the fire by the cooking line protective system and extension of flame outside the protected area of said suppression system.

On _______, the undersigned arrived at the above referenced loss site and was met by ____ State Fire Marshal _______. FM _____ gave the following account of his investigation during our meeting and, also as stated in his report, advised the undersigned of the following:

"The area of origin, to be the deep fat fryers near the drive-thru window. Examination revealed a V pat-tern at the center deep fat fryer of a three-fryer unit. The fryers were protected by an Manufacturer's dry chemical fire suppression system. The system is located within the cabinet of the fryers. Examination of the extinguisher system showed it to be cartridge operated. The model CO2 cartridge was found in the bottom of the extinguishing system cabinet. Examination of the CO2 cartridge showed the rup-ture disc to be intact and showed no signs of the fire mechanism striking the disc in anyway. The weight of the cartridge also is consistent with the cartridge being full. I examined the threads of the car-tridge and found no marks that would indicate that the cartridge had been partially installed or cross-threaded. I examined the hood and observed that the fusible link had operated. The manual activation device for this system also appeared to have been at least partially operated. I examined the agent cylinder for this system and found the rupture disc on the agent cylinder discharge line to be intact. I re-moved the fill cap from the top of the agent cylinder and found the cylinder to be full of dry chemical agent. The agent appeared to be in good condition with no packing or clumping noted. It was learned that the hood above the fryers had been cleaned on Monday evening (_____). The cleaning was conducted by John Doe Steam Cleaning. The extinguishing system is located in the cabinet of the fryers and is not readily visible making it unlikely that an employee of the restaurant had disabled the system. The condition of the extinguishing system as found is not contributory to the cause of the fire. The condition of the extinguishing system would however explain the fire growth and ability to spread un-checked." (See exhibits 5 & 6)

The suppression system was last inspected by the Fire Equipment Division of John Doe, Incorpo-rated, on ____________. (See exhibit 7)

John Doe Steam Cleaning last cleaned the cooking ventilation system of the fryers on ______. (See ex-hibit 8)

On ______, the undersigned returned to the loss site with all other interested parties. An examination and documentation of the loss site was conducted and evidence retained. The retrieved evidence along with the evidence under the security of the Fire Marshal was transferred into the custody of the undersigned. (See exhibit 9)

The area of origin consists of a three-vat,

Manufacturer's gas fryer, located under a low profile ventila-tion hood. The ventilation hood exhausted grease laden vapors through grease metal baffle type filters and out a round black pipe exhaust duct that ran from the top of the ventilation hood to an up-blast exhaust fan located on the roof. The exhaust duct had four (4) directional turns within its run to reach the roof. (See exhibit 10, Photos V-31, 41,42,43,44 & 45)

III. BASES AND REASONS FOR OPINIONS

The opinions expressed herein are the opinions that I have developed to date based on my education, experience and information I have reviewed in connection with this matter. These opinions may be re-vised if additional information becomes available to me.

1. The initial progression of the fire was direct communication of flame impingement against the grease filters of the ventilation hood for the cooking line.

2. The spread of the fire involved the burning of cooking grease that had accumulated within the venti-lation system.

3. The fire spread into the plenum area of the hood, located behind the rack of grease filters, and into the exhaust duct with great speed due to the air movement being created by the exhaust fan, located at the end of the ductwork on the roof.

4. The fire that extended into the ducts and

the failure of the _____ suppression system to extinguish the fire prevented the suppression efforts of the restaurant employees from being successful.

5. This rapid burning fire had extended into the ducts then spread out into the structural members of the building, which also prevented the fire department from quickly stopping the fire despite their im-mediate response upon notification.

6. On the basis of documents that I have reviewed and an examination of the physical evidence, it the opinion of the undersigned that the system did not activate automatical-ly or manually during this event.

7. On the basis of documents that I have reviewed in connection with this case and the physical evi-dence, Fire Equipment Division of John Doe, Incorporated failed to properly inspect the <u>Manufacturer's</u> dry chemical fire suppression system during their service of _______, by not re-installing the impellent cartridge as required by the manufacturer's manual. Mr. ______ testified in his deposition when asked:

"And on that occasion when you did your inspec-tion you did remove the cartridge of propel-lant to do your inspection, correct?"

Answer: **"I'm certain I did."** (See exhibit 19)

Again when Mr. ____, in his deposition is asked:

"Getting back to your inspection and testing that was done in _______. Do you have a specific recollection as you sit here today of reinstalling the propellant cartridge before closing the cabinet?"

Answer: **"Nor do I even recall taking it off. I mean, I don't recall, I honestly don't. But typically I would take the bottle out and put it in my tool box."** (See exhibit 20)

Mr. ______ does state that the system was inspected in accordance with the manufacturer's requirements. Again, Mr. ______ testified;

"So you're representing that you have inspected the system and that it complies with <u>Manufacturer's</u> specifications?"

Answered: **"Yes, as far as from the <u>Manufacturer's</u> certification that I got."** (See exhibit 21)

The failure to re-install the impellent cartridge during the inspection of ______, is the direct reason for the non-extinguishment of the fire while still in the protective zone of the Manufacturer's system.

Further, National Fire Protection Association Standard 96, "Ventilation Control and Fire Protection of Commercial Cooking Operations", section 11.4-4 states, **"Components of the fire suppression sys-tem <u>shall not</u> be rendered inoperable during the cleaning process."**(See exhibit 11).

To further support this hypothesis, in the dep-osition of _______., an employee of John Doe Steam Cleaning, who assisted with the last cleaning of the Loss Location two days before the fire, he testified as follows:

"Have you ever disconnected an automatic fire suppression system--?"

Answered, **"No."** (See exhibit 22)

Again when asked:

"Have you ever observed any of the other employees of John Doe Steam Cleaning discon-nect a fire suppression system?"

Answered, **"No."** (See exhibit 22)

8. On the basis of documents that I have reviewed in connection with this case and the physical evi-dence, Fire Equipment Division of John Doe, Incorporated failed to properly inspect the Manufacturer's dry chemical fire suppression system during their service of _______, by not properly inspecting the duct-work of the ventilation system during their inspections of the Manufacturer's Suppression System.
The Manufacturer's Manual for the____ Fire Suppression System, front cover states, **"Fire extin-guishers are a mechanical device. They require periodic care. Maintenance is a vital step in support of your system. As such it**

should be performed in accordance with NFPA 96 by a qualified Manufacturer's service man. Maintenance to provide maximum assurance that your fire extinguishing system will operate effectively and safely should be conducted at six-month in-tervals. Inspection to provide reasonable assurance that your fire extinguishing system is charged and operable should be conducted at more frequent intervals." (See exhibit 12)

NFPA 96, Chapter 1, 1-1.2 states, "The provisions of this standard are considered necessary to pro-vide a reasonable level of protection from loss of life and property from explosion or fire. They reflect conditions and the state of the art prevalent at the time the standard was issued. **This standard is intended to be applied as a united whole. It cannot provide safe design and operation if parts of it are not enforced or are arbitrarily deleted in any application."** (See exhibit 13)

Underwriters Laboratories and Ansul Manufacturing require that NFPA 96 be adhered to in its entirety so that the listing of the system is not violated.

NFPA 96, Section 4.2.1 states, **"Where enclosures are not required, hoods, grease removal de-vices, exhaust fans, and ducts shall have a clearance of at least 457mm (18 in.) to combustible material."** (See exhibit 14).

The exhaust duct run was against the combustible members of the building. This allowed for the ensu-ing fire within the ductwork to escape out and to the combustible structure of the building. (See exhibit 15)

NFPA 96, Section 7.5.2.1 states, "**All seams, joints, penetrations, and duct-to-hood collar connec-tions shall have a liquidtight continu-ous external weld.**" (See exhibit 18)

No liquid tight welds existed on the duct run. All connections were simple slip-on joints. This, again, allowed the fire to progress from the interior of the duct out to the combustible structure of the build-ing.

9. On the basis of documents that I have reviewed in connection with this case and the physical evi-dence, John Doe Steam Cleaning failed to properly clean the exhaust duct of the ventilation system during their cleaning just two (2) days before the fire loss by not cleaning the ventilation system to bare metal due to the lack of access panels.

NFPA 96, Section 7.3.1 states, "**Openings shall be provided at the sides or at the top of the duct, whichever is more accessible, and at the change of direction.**" (See exhibit 16)

Access panels did not exist at all changes of direction. This lack of access panels would not allow for all accumulated grease to be removed from the ventilation system.

10. On the basis of documents that I have reviewed in connection with this case and the physical evi-dence, John Doe Steam Cleaning failed to properly clean the exhaust duct of the ventilation system during their cleaning

just two (2) days before the fire loss by not properly re-installing the existing ac-ess panels.

NFPA 96, Section 7.4.3.2 states, "**Access panels shall have a gasket or sealant that is rated for 815.6C (1500 F) and shall be grease tight.**" (See exhibit 17)

No gaskets exist on the access panels, thus allowing for fire to escape outside the ductwork and ignite the combustible members of the building.

Further, had the Manufacturer's system been properly inspected, had the ductwork clearance viola-tions been brought to the attention of the Loss Location so that corrective action could be taken and had the proper gaskets been installed after the cleaning of the ventilation system, this fire would have been extinguished while still encapsulated within the ventilation system of the cooking line.

The opinions contained in this report are based on personal observations at the fire scene; statements of witnesses and other documents reviewed as of the date on this report and were developed to a reasonable degree of certainty. The author agrees to a reconsideration of the conclusion if new evidence becomes available. This investigation was conducted using NFPA 921 as a guide and other authoritative sources.

Notes

Notes

Notes

Notes

Notes